Ecologies of Socialisms

German Life and Civilization

Vol. 70

PETER LANG
Oxford · Bern · Berlin · Bruxelles · New York · Wien

Sabine Mödersheim, Scott Moranda,
and Eli Rubin (eds)

Ecologies of Socialisms

Germany, Nature, and the Left in History, Politics, and Culture

PETER LANG
Oxford · Bern · Berlin · Bruxelles · New York · Wien

Bibliographic information published by Die Deutsche Nationalbibliothek
Die Deutsche Nationalbibliothek lists this publication in the Deutsche Nationalbibliografie; detailed bibliographic data is available on the Internet at http://dnb.d-nb.de.

A catalogue record for this book is available from the British Library.

Library of Congress Cataloging-in-Publication Data

Names: Mödersheim, Sabine, 1959- editor. | Moranda, Scott, 1975- editor. | Rubin, Eli, 1975- editor.
Title: Ecologies of socialisms : Germany, nature, and the left in history, politics and culture / Sabine Mödersheim, Scott Moranda, Eli Rubin (eds).
Description: Oxford ; New York : Peter Lang, [2019] | Series: German life and civilization ; 70 | Includes bibliographical references and index.
Identifiers: LCCN 2018050511 | ISBN 9781787075771 (alk. paper)
Subjects: LCSH: Socialism--Germany--History. |
Environmentalism--Germany--History. | Communism and ecology--Germany--History.
Classification: LCC HX274 .E36 2018 | DDC 320.53/10943--dc23
LC record available at https://lccn.loc.gov/2018050511

Cover image: Demonstration against the planned nuclear power plant at Wyhl. Photograph by Leo Horlacher © Archiv Soziale Bewegungen. Reproduced with permission.

ISSN 0899-9899
ISBN 978-1-78707-577-1 (print) • ISBN 978-1-78707-579-5 (ePub)
ISBN 978-1-78707-580-1 (mobi) • ISBN 978-1-78707-578-8 (ePDF)

Published by Peter Lang Ltd, International Academic Publishers,
52 St Giles, Oxford, OX1 3LU, United Kingdom
oxford@peterlang.com, www.peterlang.com

This publication has been peer reviewed.

Contents

Figures

Acknowledgments

This book benefitted from the hard work and dedication of many people. The editors would like to thank first and foremost the German Studies Association. The GSA instituted "interdisciplinary networks" some years ago, as part of its commitment to interdisciplinary German studies. This led to the formation of the Environmental Studies network, and more recently, the German Socialisms network. The GSA also encouraged cooperation *between* interdisciplinary networks, and this led to a collaboration between the German Socialisms network and the Environmental Studies network. A huge thank you goes to Marc Silberman, who encouraged both the formation of the German Socialisms network, and the cooperation between German Socialisms and Environmental Studies. Also a huge thank you is owed to April Eisman and Benjamin Robinson, who, along with Eli Rubin, helped found the German Socialisms network and oversaw this collaboration with Environmental Studies, led by Scott Moranda and Sabine Mödersheim. Tom Lekan and Katharina Gerstenberger deserve great thanks for helping launch the Environmental Studies network and bringing Sabine and Scott together to continue its work.

Another enormous debt of gratitude belongs to our editor at Peter Lang, Laurel Plapp. It was she who had the idea of publishing the results of our networks' collaboration. Her encouragement, steady guidance, and patience at our somewhat slow pace meant everything to the successful outcome of this book! And we would be remiss in not thanking Jost Hermand, the editor of this series, who supported this project from the beginning, and whose own work defined the careers and the work of all three of the editors and many of the contributors of this volume. Thanks as well to our anonymous reader, who helped us make many improvements. Thanks also to Simon Phillimore at Peter Lang who helped with the editorial process.

Most of all, we wish to thank our amazing contributors, all of whom are engaged in work of the highest importance, and producing cutting edge and fascinating scholarship. The Anthropocene is here, and we must grapple with it in all its dimensions, as these scholars are doing. With their hard work and excellent research and writing they are helping lead the way for many future scholars.

ELI RUBIN AND SCOTT MORANDA

Introduction

Is Germany the "Greenest nation" as Frank Uekötter has provocatively asked?[1] Forget the Berlin Wall, forget the Cold War, World War II, the Nazis, the Holocaust, the *Kaiserreich*, and forget the storied history of the Social Democratic Party – the SPD. Germany is now the nation of separated trash, solar panels and wind power. The only *Wende* people talk about now, in relation to Germany, is the *Energiewende*. And perhaps, that is part of the point. The *Energiewende* provides for a national identity that generates warm, positive press and seemingly avoids the more divisive and troubling aspects of past markers of Germanness.

Even if the reality of the "greenest nation" is more complicated, it is undeniable that Germany has been at the center of the global history of environmentalism. Scientific conservation had deep roots among the foresters, urban planners, and other technical experts of Wilhelmine Germany who had outsized influence on Progressive reformers in the United States and British civil servants across their Empire.[2] German-speakers also shaped Romanticism, ecology, and notions of holistic interconnectedness, most importantly in the figure of Alexander von Humboldt, whose *Kosmos* had a global influence.[3]

The rise of modern environmentalism also was central to postwar German history. Beginning in the early 1970s with Earth Day and the

1 Uekötter, *The Greenest Nation? A New History of German Environmentalism* (Cambridge, MA: MIT Press, 2014).

2 S. Ravi Rajan, *Modernizing Nature Forestry and Imperial Eco-Development 1800–1950* (Oxford: Oxford University Press, 2010); Daniel T. Rodgers, *Atlantic Crossings: Social Politics in a Progressive Age* (Cambridge, MA: Harvard University Press, 2009).

3 Andrea Wulf, *The Invention of Nature: Alexander von Humboldt's New World* (New York: Vintage Books, 2016).

anti-nuclear movement, and fueled by massive protests and occupations in places such as Wyhl and Brokdorf, the environmental movement coalesced into a new political party – the Green Party – in 1980. And the Greens quickly ascended from a ragtag collection of activists to a political party holding seats in the Bundestag and influencing the national debate and policy. In 1983, *Der Spiegel* reacted to the growing movement by publishing an issue devoted to the supposed dying-off of German forests due to acid rain, thus igniting the *Waldsterbendebatte* [the forest death debate].[4] The *Waldsterbendebatte* transcended political and ideological boundaries – everyone, Right, Left, or Other, was scared and thought Germany's forests were going extinct. The center-right Kohl government began issuing environmental regulations to limit industrial pollution, and even though the Greens were not part of the ruling coalition, the legislation that Bonn produced went beyond anything the Greens could have dared to hope for. From that moment on, Germany was on track to become the "Greenest nation," a path it has followed until the current era of the *Energiewende*.

At the same time, the rise of the Green Party, and the modern environmental movement, represented a transcendence of, a departure from, and even a repudiation of, traditional Left politics. As Eli Rubin discusses in his contribution to this volume, many of the founders and eventual leaders of the environmental movement and the Green Party were originally Leftists, such as West Germans Joschka Fischer and Otto Schily, or East German Rudi Bahro, who came to enjoy a popularity and level of influence as Greens they never had as Reds. By the 1970s and early 1980s, many left-wing German intellectuals, who had already begun drifting from "traditional" Marxism, came to see the exploitation of *nature* by humans as more central in their analyses and critiques than the exploitation of *humans* by humans. Carl Amery, for example, argued that "ecology" should replace "economy" as the central object of politics.[5] The result was that the Green

4 Tobias Huff, *Natur und Industrie im Sozialismus: Eine Umweltgeschichte der DDR* (Göttingen: Vandenhoeck & Ruprecht, 2015), 9 & 276.

5 Andrei Markovits and Philip Gorski, *The German Left: Red, Green and Beyond* (New York: Oxford, 1993), 127.

Party, and environmentalism more broadly, became more appealing to people from a diverse range of political backgrounds, not just the Left. In particular, elements of the *völkische* Right were attracted to environmentalism as well, as Rubin explains.

What tied them together was a critique of *growth*. Growth was a broad concept that meant different things to different people, but for many disaffected on the Left and the Right, it meant broadly the expansion of the human population, and the expansion of the economy to serve the needs *and the wants* of that population. Growth meant, to Green activists in Germany and beyond, an endlessly expanding population with endlessly expanding desires – for modern housing, roads, clothes, toys, entertainment; in short, the "good life." And, of course, more consumer goods and modern amenities meant more consumption of natural resources, and it also meant more population growth, which meant more needs and desires, and so on until the inevitable exhaustion of all resources and collapse of the economy, and ultimately the human race. And once the world ended, it would not matter who was a "capitalist" or a "communist."

Yet, for the communist parties and their leaders who were engaged in building communism in Eastern Europe, the pressure to grow the economy and to produce the good life for their citizens was a question of self-preservation because of their competition with the capitalist world. For the GDR, especially after Erich Honecker's 1971 ascension and announcement of the "Unity of Economic and Social Policy" (also referred to as "real existing socialism"), this meant that almost *everything* was subordinated to the demands of the socialist consumer society – to growth. For many East German intellectuals, who still considered themselves Marxists, like Rudi Bahro or Wolfgang Harich, it was a bridge they could not cross. Harich, in his 1975 *Kommunismus ohne Wachstum* [Communism without Growth] estimated that 70 percent of all industrial products were "superfluous" – and this was in the GDR in 1975 (imagine him in a suburban outlet mall in twenty-first-century America).[6] The SED responded to his ideas by

6 Hermand, *Grüne Utopien in Deutschland: Zur Geschichte des ökologischen Bewußtseins* (Frankfurt/Main: Fischer, 1991), 144.

proclaiming that "communism without growth would not be some other kind of communism, but no communism at all."[7]

The toll that the SED's relentless pro-growth policies took on the environment in East Germany is well known now, and was hard to ignore at the time for East Germans as well. The immediacy of the pollution caused by communism's supposed reliance on growth simply trumped any theoretical arguments about the long term benefits that would outweigh that pollution. And it was this fact that gave momentum to the grassroots environmental groups who helped spark the protests that led to the fall of socialism on German soil – the *Wende* of 1989–90. Many of these groups, furthermore, had help from West German Green politicians and activists. So, although many tend to associate environmentalism with the Left, in 1989–90, it was the "Greens" who were opposed to the "Reds," and the reason was that for many Greens, it was *socialist* East Germany (and by extension the entire Soviet world) that became the face of prometheanism – the modern impulse to conquer and harness nature – more than the West. What meaning did "Left" or "Right," "Green" or "Red" have any longer?

If history writing is essential to nation building and if Germany has reimagined itself as the greenest nation, the "Left" thus risks being written out of reimagined nation's origin story. While socialism remained tightly associated with prometheanism, Germans (and Europeans generally) have worked to build a "light green" society where consumer choice and sustainability could go hand-in-hand. The dream of a green capitalism fueled a movement to save the planet through the market. Reformers pleaded with consumers to make greener shopping choices, and regulation often worked through market incentives. Helmut Kohl's project to make the former East Germany blossom again captured the spirit of this age, with its vision of a revitalized landscape under a capitalist economy.[8]

7 Huff, *Natur und Industrie im Sozialismus*, 201.

8 On light-green society, see Michael Bess, *The Light-Green Society: Ecology and Technological Modernity in France, 1960–2000* (Chicago: University of Chicago Press, 2003) and Sandra Chaney, *Nature of the Miracle Years: Conservation in West Germany, 1945–1975* (New York: Berghahn Books, 2012).

If one wanted to find them, the roots of the "greenest nation" can be found in any of the major epochs of the German past, even the most horrific. The romantic poets and the *Turner* movement looked to the natural world as the wellsprings of early nineteenth-century German nationalism. The forest, in particular, became celebrated as the refuge for a particular German spirit in these burgeoning nationalist discourses.[9] Later in the nineteenth century, "homeland" (*Heimat*) preservation societies and tourist activities cemented the notion of a nation cobbled together from many different "homelands." The preservation of distinctive regional landscapes became essential to cultural survival.[10] Landscape architects and urban reformers in the early twentieth century hoped to cure the ills of modern industrial society by more rationally planning the built environment to incorporate healthy green spaces and to tame the excesses of capital.[11] In the twentieth century, visions of an organic modernity shaped the hopes of many Germans, some of whom saw in Nazism the opportunity to forge that alternative modernity. Natural forests, highways nestled naturally into the landscape, biodynamic farming were all potential pathways considered during the Third Reich.[12] Fears of plunder by occupation forces after the war kept the landscape, nature, and the environment in the minds of West Germans, as they rallied together

9 Jeffrey K. Wilson, *German Forest: Nature, Identity, and the Contestation of a National Symbol 1871–1914* (Toronto: University of Toronto Press, 2016).

10 Alon Confino, *The Nation as a Local Metaphor: Württemberg, Imperial Germany, and National Memory, 1871–1918* (Chapel Hill: University of North Carolina Press, 1997); Celia Applegate, *A Nation of Provincials: The German Idea of Heimat* (Berkeley: University of California Press, 1990); William H. Rollins, *A Greener Vision of Home: Cultural Politics and Environmental Reform in the German Heimatschutz Movement, 1904–1918* (Ann Arbor: University of Michigan Press, 2000).

11 Thomas M. Lekan, *Imagining the Nation in Nature: Landscape Preservation and German Identity, 1885–1945* (Cambridge, MA: Harvard University Press, 2004).

12 Franz-Josef Brüggemeier, ed., *How Green Were the Nazis? Nature, Environment, and Nation in the Third Reich* (Athens: Ohio University Press, 2007); Frank Uekötter, *The Green and the Brown: A History of Conservation in Nazi Germany* (Burnaby, BC: University of Simon Fraser Library, 2006).

to protect the German forest and landscape from what they feared would be Anglo-American harm.[13]

Arguably, the German Democratic Republic has been left out of that history altogether. In part, notions of the Left's prometheanism has been cemented by images of environmental protest and ecological destruction in the waning years of the GDR. Histories of the working-class movement largely focus on urban settings and problems. The historiography of the GDR generally has paid little attention to environment, and when it has, it largely reinforces the narrative of prometheanism. Histories of the GDR that reveal a richer, more complicated relationship with the land make little headway against this common perception. Nonetheless, the dictatorship, and the workers' movement before that, did not just kill trees and poison waters. Individuals and groups in the government and in mass organizations worked (to varying degrees of success) to create a "socialist Heimat," to provide camping vacations to workers, to add fresh air and sunlight to housing projects, to maintain hiking trails, and to beautify villages.[14]

Recent environmental histories of the Cold War have also reminded us to avoid oversimplifications when it comes to the ecologies of socialism. The battle to provide "greater living standards" demanded greater exploitation of natural resources on both sides of the Iron Curtain. While ideologies certainly shaped the choices made by both sides, strategic planning, military preparedness, and economic competition often mattered more. Older systems of colonialism, economic protectionism, or nationalism often remained just as important as new, particular Cold War concerns in shaping daily life and land use.[15] Cold War tensions not only led

13 Scott Moranda, "It Makes No Sense to Work against Nature: Cold War Modernization in West German Agriculture," in *Nature and the Iron Curtain*, eds. Astrid Mignon and John McNeill (Pittsburgh, PA: University of Pittsburgh Press, 2019).

14 Jan Palmowski, *Inventing a Socialist Nation: Heimat and the Politics of Everyday Life in the GDR, 1945–1990* (Cambridge: University of Cambridge Press, 2013); Scott Moranda, *The People's Own Landscape: Nature, Tourism and Dictatorship in East Germany* (Ann Arbor: University of Michigan Press, 2014); Huff, *Natur und Industrie im Sozialismus*.

15 John R. McNeill and Corinna R. Unger, *Environmental Histories of the Cold War* (Cambridge: University of Cambridge Press, 2013); Kate Brown, *Plutopia: Nuclear*

to degradation but also shaped the mentalities that gave rise to modern environmentalism, as visions of nuclear apocalypse or global confrontation shaped scientific conservation and, later, nonconformist rejections of consumerism and growth.[16]

***

So, what is environmentalism? Is it Left, Right – both or neither? Where do the Greens fit in the ideological and cultural spectrum of Germany, and more broadly, everywhere?

The question, it turns out, is complicated and not an easy one to answer. Yet, for that complexity, exploring the complicated relationship between left-wing thought and movements and ecology opens doors onto a new understanding of both "red" and "green" strands of German history and culture, as well as the history and culture of societies beyond Germany. The aim of this volume is to explore the relationship of socialism and environmental or ecological thought and movements – and to do so from an interdisciplinary and transnational perspective – including fields of history, literature, ethnology and urban studies, and ranging from East Germany and West Germany to Austria and even Vietnam.

One of the themes that emerges from these works, very quickly, is that there is no monolith of "socialism" nor of "ecology." Nor is the relationship between socialism and ecology simple to explain. There were, from the beginning, a multiplicity of socialisms in Germany, and this continued throughout the Cold War, including within the GDR itself. As Scott Moranda, Tobias Huff, and Astrid Mignon Kirchhof describe in their essays, there were in fact numerous other paths that the socialist GDR could have taken in regards to its treatment of the landscape and nature. Moranda writes about the landscape planner Georg Bela Pniower who

---

Families, Atomic Cities, and the Great Soviet and American Plutonium Disasters (Oxford: Oxford University Press, 2015); Thomas Robertson, *The Malthusian Moment: Global Population Growth and the Birth of American Environmentalism* (New Brunswick, NJ: Rutgers University Press, 2012).

16 Jacob Darwin Hamblin, *Arming Mother Nature: The Birth of Catastrophic Environmentalism* (Oxford: Oxford University Press, 2017).

initially advocated in East Germany for a vision of socialist agriculture that was sustainable, and which would overcome the "metabolic rift" – the rupture of human needs from the natural carrying capacity of the land through urbanization, industrialized agriculture and distribution networks, all of which were caused by capitalism. Pniower argued for a gradual transition from smaller to larger farms, always taking care not to exacerbate the metabolic rift. As Moranda writes, Marx himself was deeply concerned about soil depletion and erosion, and saw *Raubbau* – plunder, of the natural environment – as one of the truly damaging effects of capitalism. The essay seeks to identify links between nineteenth-century Marxist thought about metabolism and agriculture to the early days of East German agriculture when policy was still up in the air.

Huff writes about other East German landscape planners such as Reinhold Lingner who, like Pniower, did not see any inherent conflict between socialism and the ecology – in fact, inspired by Marx's ideas, Lingner believed it was socialism that promoted a healthier relationship between humans and nature, because it was capitalism that promoted the "plunder" of the natural environment, and that socialism ultimately would lead to a healthier balance between humans and the ecology. It was Lingner who wrote the East German plan for the transformation of nature, which Huff notes was not the apex of Stalinist gigantomania that many assumed it to be. And Kirchhof writes about Erna and Kurt Kretschmann, East Germans who lived the values of sustainability and harmony with nature which they wanted to see reflected in their broader society. Unable to get the Politburo to adopt the kind of ecological socialism in which they believed, they instead created a "counterworld" as Kirchhof writes, in which they quietly welcomed visitors to their small "House of Nature Care," in which they taught workshops in sustainable living. Over the course of three decades, 80,000 people passed through their doors as visitors, friends and enthusiasts hoping to learn. The Kretschmanns created a niche and planted seeds – in a literal and in a figurative sense. The "seeds" they planted in their "niche" grew into the "grassroots" that did eventually overwhelm what had seemed to be the powerful Moloch of the growth and production-driven East German political economy.

In all these cases, we read about ideas and efforts that seemed invisible, or perhaps paths not taken. Though they are left out of the standard narrative about a promethean socialism exploiting nature, they play an important role once we see these different strands of ecological belief and thought as existing within an ecology of socialism. In fact, the very word "ecology" is important to explore here, because it has deep roots in German history and culture. The term was coined by scientist and philosopher Ernst Haeckel in 1866 in his groundbreaking *Generelle Morphologie der Organismen* [General Morphology of Organisms]. A naturalist by training, Haeckel was deeply influenced both by Darwin, as well as by Goethe – in particular, Goethe's rejection of Cartesian dualism, and what he saw as the resulting promethean or "mechanistic" view of nature as something to be "dismembered and deconstructed" (*zerlegen und zerstückeln*) as Jost Hermand has described.[17] Goethe believed that science and technology, seen as a tool for humans to manipulate nature, disrupted what he believed to be the "inner unity" of all matter, and which would eventually lead to the death of all of nature.[18] Haeckel was also deeply influenced by Spinoza, Bruno and especially other German thinkers of the nineteenth century who were seeking to articulate some kind of monist philosophy, including Gustav Fechner, Carl Vogt, and Ludwig Büchner. All of these were influenced to some degree by the ancient Greeks, including the Ionian philosophers, Empedocles, Epicurus, and Theophrastus, all of whom believed in a kind of hylozoism – the idea that nature, including humans, is a unity, and that consciousness or "soul" exists in non-human objects as it does in humans.[19] Haeckel's hylozoism stopped short of claiming, as Fechner and Goethe did, that consciousness itself could be found in every bit of matter, but his monism was almost just as radical – all matter, he believed, had "sensation" and humans were just concentrations of that which was already in nature.[20] It was this sense of the interconnectedness of all matter, humans and non-humans, that

17 Hermand *Grüne Utopien*, 50.
18 Hermand, *Grüne Utopien*, 51.
19 David DeGrood, *Haeckel's Theory of the Unity of Nature* (Boston, MA: The Christopher Publishing House, 1965).
20 DeGrood, *Haeckel's Theory*, 44–45.

motivated him to utilize the term coined by Theophrastus, *oikeios*, which had originally meant something like "niche" – Theophrastus wanted to convey the fact that organisms are adapted to their environments, and cannot be simply taken out of their natural habitats to serve any human purpose.[21] For Haeckel, "ecology" meant more than just "niche" – it meant the interconnectedness of all living and non-living things, a definition much broader than the conventional view of "ecology" as representing non-human life-forms (flora and fauna).

Across the decades, and centuries, and through Germany (and beyond) at the end of the nineteenth century, there was an ecology of ideas about ecology – an ecology of ecologies. This ecology of ecologies profoundly shaped German history. The "Monist League" that Haeckel created to advance his views had an enormous impact, and drew together a whole interconnected network of groups, beliefs and ideologies that had their own take on *ecology* as a system of interconnection in which one part influenced all the others and vice-versa. Rudolf Steiner, the founder of anthroposophism, for example, was also very much influenced by the same Goethean monism as Haeckel, and the entire *Lebensreform* movement was deeply influenced by Haeckel's idea of the "ecology," representing an attempt to put holism into actual praxis.[22]

Even groups less influenced by "monism" drew on influences from Goethe and Humboldt to explore interconnectedness. German geographers looked to identify cultural "isotherms" denoting unique cultural communities. For some geographers, this translated into the mapping of racial spaces tying human populations to certain environmental conditions or dooming other cultures to extinction as their unique habitats fell to European expansion. Landscape designers imagined the countryside as a living collection of distinct forms that, when kept in balance, sustained the "health" of a place. Social hygienists, geographers, nature preservationists, anthropologists, and lifestyle reformers shared vocabulary about systems and wholes and borrowed from each other, whether they sat on the right or left of the political

21 J. Donald Hughes, "Theophrastus as Ecologist" *Environmental Review: ER* 9, no. 4 (Special Issue: "Roots of Ecological Thought"; Winter, 1985): 296.

22 Hermand, *Grüne Utopien*, 92–95.

spectrum. Trends like *Lebensreform* had long afterlives. For example, they re-emerged in the form of the Kretschmanns, and the anthroposophists played an important role in the shaping of Green politics and ideology in the 1970s and 1980s, as prominent founders of the movement and former Marxists like Rudi Bahro and Rainer Langhahns became, like so many Germans almost a century earlier, *Aussteiger* [dropouts], joining communes.

This volume thus explores socialism's ecology in two different ways. On the one hand, by ecology we refer to the real biophysical limits that bind a natural system. Socialist societies and ideas exist within those material realities, and the choices made by socialists can sustain or disrupt natural communities. In this case, we are concerned with how aware socialist actors are or were of those biophysical limits and how they did or did not respond to them. In this sense, it is important to remember that socialist states in Cold War eastern Europe were cut off from most global material flows. While West Germany imported energy and nutrients in an American-dominated economic sphere of relatively open markets and also continued to benefit from colonial and neocolonial trade inequities, East Germany remained much more self-reliant. To a certain degree, the East even had to give up some natural resources to its hegemon to the East. While some historians have acknowledged this wider context for the environmental histories of the two Germanies in the twentieth century, they remain the exception.[23]

On the other hand, we explore a more metaphorical meaning of ecology. Where did socialism fit in the interconnected world of ideas and movements in the German-speaking world? Such an approach might make some readers uncomfortable given the stakes of our planet's current ecological predicament. Our project is not meant to distract from the larger questions of ecological survival, but rather is meant to re-insert socialism into our stories and histories about the human place in the natural world and about the "greenest nation." Socialist thinkers were not isolated from the larger intellectual ecosystem of central Europe, and the questions of

23 Raymond Dominick early on acknowledged how the two Germanies occupied very different global systems. See his "Capitalism, Communism, and Environmental Protection: Lessons from the German Experience" *Environmental History* 3, no. 3 (1998): 311–332.

metabolism, interconnectedness, and whole systems also influenced left-wing ideas. Of course, German Marxists vocally condemned ideas about the human-nature relationship that they believed to be too tinged with Malthusian panic, but this did not mean they were blind to the natural systems within which human societies lived.

Haeckel was not the only history-altering thinker to publish a magnum opus in the 1860s, of course; Karl Marx's *Capital* was published just one year after *Generelle Morphologie* and Marx was very much preoccupied with many of the same questions as Haeckel, in terms of the relationship between humans and nature, especially in regards to agriculture and soil depletion. As noted above, Marx was very concerned with the metabolic rift, and in fact was so concerned that he put the concept of the "metabolism" (*Stoffwechsel*) between humans and non-human nature at the heart of his concept of labor, as John Bellamy Foster has written.[24] Yet, Marxian notions of "metabolism" and "metabolic rift" were largely forgotten for many decades especially in the West, where Socialism – and Marx himself – came to be considered as promethean. Neil Smith, in particular, has argued that Marx's work has been profoundly misinterpreted, in part by theorists associated with the "Frankfurt school," in particular Alfred Schmidt, for whom Marx's work represented dualism, and was therefore responsible for the worst of human subjects dominating the objects of the natural world. In reality, argues Smith (and other scholars such as Foster and Moishe Postone), Marx saw a dialectical *unity* of human and nature, not a *duality*. Smith goes further and accuses the Frankfurt School, in particular Schmidt and Marcuse, of actually being guilty of dualism and thus replicating a bourgeois view of nature.[25] It should be noted that Smith's critique does not extend to all Frankfurt School theorists

24 Foster, "Marx's Theory of Metabolic Rift: Classical Foundations for Environmental Sociology" in *American Journal of Sociology* 105, no. 2 (September 1999): 380.

25 Smith, *Uneven Development: Nature, Capital, and the Production of Space* (Oxford: Oxford University Press, 1984), 22–23. Also see Foster, *Marx's Ecology: Materialism and Nature* (New York: Monthly Review Press, 2000), 165 and Postone, *Time, Labor, and Social Domination: A Reinterpretation of Marx's Critical Theory* (Cambridge: Cambridge University Press, 1993), 311–312. Thanks to the anonymous reader and Deborah Cook for these particular citations.

– as Deborah Cook has recently shown, Theodor Adorno was not a dualist and did not see Marx as promethean either.[26]

The stakes of this rather arcane sounding debate are actually quite high. A dualistic worldview, seeing humans as distinct from non-human nature, is not truly an ecological one, at least not in the original sense intended by Haeckel (or many of the ancient Greeks, for that matter). As Jason Moore, in his groundbreaking recent work *Capitalism in the Web of Life* argues, dualism has come to inform most of environmental, or "Green," thought, activism, and policy, and as a result, it is often couched in terms of what humans "do to" "the environment." Thus, it loses track of the role that capitalism as a system itself plays in shaping both the environment and at the same time human society. In particular, capitalism is responsible for creating dualistic modes of thought that hide the essential interconnectedness or "metabolism" that binds humans and non-humans. To transcend such thought, Moore argues, we need to return to the original idea of ecology – the *oikeios* – and to begin seeing human and non-human actors as part of what he calls a "double internality," capitalism-in-nature and nature-in-capitalism, which is essentially his eponymous "web of life." Referencing Marx, Hegel, Haeckel, Theophrastus and Heraclitus, Moore writes:

> Is the image of nature as passive mud and dirt – a place where one leaves a footprint – really the best metaphor to capture the vitality of the web of life? I think we can do better ...
>
> Instead of a contemporary world produced by two discrete, interacting, substances – Society and Nature – we might instead look at the history of modernity as co-produced, *all the way down and through*. One substance, Humanity, does not co-produce historical change with another substance, Nature. Rather, the species-specificity of humans is already co-produced within the web of life. Everything that humans do is a *flow of flows* in which the rest of nature is always moving through us.[27]

26 Cook, *Adorno on Nature* (London: Routledge, 2011), 25–26. Cook also mentions that Schmidt recognized the centrality of "metabolism" in Marx as well, contradicting somewhat Neil Smith's criticism of Schmidt.

27 Moore, *Capitalism in the Web of Life: Ecology and the Accumulation of Capital* (New York: Verso, 2015), 6–7. Our emphasis.

There is, therefore, more to the term "ecology" than soil measurements, nature centers, or sustainable gardening. In world history, but also very much in German history, it is a term that implies both a kind of interconnectedness, with elements of monism and dialectics, in which nothing can really be taken out of context, and all parts of the system impact all other parts, so there is no clear center, but rather networks of networks, or flows of flows, or – ecologies of ecologies; *and* it implies a reaction to industrialization and mechanization brought about by capitalism's unquenchable thirst for new sources of accumulation. Ecological thought therefore should be sensitive both to the dialectical *Wechselwirkung* of capitalism, and socialism, and the non-human environment – and, it should be able to think through many nodes at once, without privileging any one. In particular, any investigation of ecology and socialism needs to be thought through ecologically in this way.

While Moore's accusations of dualism (especially when leveled against Marxist environmentalists) can be overwrought, his conceptualization of the capitalocene remind us to remain alert to flows, dead ends, shifting grounds, and other ways that natural material and ideas pass along contours constantly being reworked by ideas, systems, and natural phenomena.[28] While this volume does not seek to rewrite the environmental history of East Germany or other socialist states, it insists on the value of taking socialists seriously in our histories of environmental thought. Socialist planners and thinkers adapted older German ways of living in and thinking about nature to their material and ideological realities. Many of those adaptions proved dead ends, but unearthing them in this volume might reinsert them into our narratives, whether of German-speaking Europe, the anthropocene, or the capitalocene. Solutions to our global ecological crisis will need to acknowledge the promethean threat to our ecosystem,

28 "Two Views on Marxist Ecology and Jason W. Moore," (June 23, 2016) on Climate and Capitalism (website), ed. Ian Angus, <https://climateandcapitalism.com/2016/06/23/two-views-on-marxist-ecology-and-jason-w-moore/>. For a defense of Moore, see Louis Proyect, accessed December 29, 2018, <https://louisproyect.org/2016/10/10/jason-moores-capitalism-in-the-web-of-life-ecology-and-the-accumulation-of-capital/>.

but their effectiveness likely will require a sensitivity to the constant shifting and entangled structure of that threat. Seeking out and defending an untouched nature likely will be impossible as nature unsullied by capital flows will be hard to find. Given how profoundly our planet is shaped by capital, older socialist critiques of capitalism might provide insights relevant to today's predicament. In fact, if socialist thinkers today want to fashion responses to climate change, acknowledging and revisiting the history of socialist engagement with ecology will be vital.

***

The works in this volume were brought together under intentionally interdisciplinary auspices. They originated as papers in a series of panels at the German Studies Association Annual Conference in 2016, in San Diego, which were arranged by two of the GSA's "interdisciplinary networks" – the "German Socialisms" network and the "Environmental Studies" network. The papers presented encompassed a wide range of disciplines and methodologies, but all were committed to thinking about the paradoxes, forgotten pathways, dialectical twists and turns that surround how ecology, the Left, capitalism and the Cold War have played out in Germany and beyond since 1945. In each case, the notion of ecology as "nature" and ecology as a more abstract sense of distributed interconnectedness exist parallel to each other, but also collapse into each other: using ecology to discuss capitalism, communism, and postcommunism opens up ecological ways of understanding those same topics that can move our thinking forward.

So, for example, Julia E. Ault writes about water – a highly timely subject, as it has become the center of environmental activism and broader critiques of neoliberal governance and the relationship between space, capitalism, race and the exact kind of "double internality" that Moore describes, for example, in the case of the poisoning of the water supply in Flint, Michigan. For Ault, rivers formed borders but because they were always in flux – a flow of flows – they also destabilized borders. Increased pollution in East German rivers flowed into its neighbors' watersheds and water supplies. Water policy forced East Germany to compete with its neighbors when it was supposed to be cooperating with them, and because waterways form their own ecology, and because humans rely on water for life, by necessity water as an ecology created another ecology, a political ecology of socialism, as it joined and also brought into conflict the various neighbors of socialist Eastern Europe.

Christine Schwenkel writes about a prefabricated housing development, designed and built by East Germans, in the Vietnamese city of Vinh, as part of the postwar reconstruction in the late 1970s. The Vinh settlement was built in a very similar way as the massive prefabricated housing block settlements being constructed at the same time in East Germany as part of the ambitious "Housing Program," a centerpiece of the Honecker regime's "real existing socialism." In East Germany, in places such as Berlin-Marzahn, Leipzig-Grünau, or Karl-Marx-City Fritz-Heckert, hundreds of thousands of apartments, for millions of East Germans, were built in assembly-line fashion out of prefabricated panels, or *Platte* (the provenance of the name *Plattenbausiedlungen*) on former farmlands and small villages on the outskirts of cities. The Housing Program was very much what critics like Rudi Bahro or Wolfgang Harich had in mind when they railed against a socialism that wanted to just keep bulldozing, building, producing, polluting. It is very much what outside critics, including those from the West German new Left, had in mind when they pushed an image of East German socialism as "promethean" or even guilty of "gigantomania."[29]

And yet, things were not so simple. In Vietnam, Schwenkel writes, East German planners took nature very much into account, building the apartment blocks at 23 degree angles to maximize airflow through them, and integrating trees as a vital part not just of a high quality of life, but as something central to Vietnamese traditional culture. And indeed, the socialist building of massive housing settlements was a paradox: on the one hand, it did represent the promethean arrogance of remolding and mastering nature, tearing it apart (*zerstückelt und zerlegt* à la Goethe) and rebuilding to suit human needs, or wants, which were the central focus of "real existing socialism." But at the same time, and following Corbusian precepts,

29 See Eli Rubin's contribution in this volume, in particular referencing David Blackbourn's use of the term "gigantomania."

the *Plattenbausiedlungen* of East Germany were built to be verdant,[30] and today, they are, with urban gardens, urban forests, and remarkable amounts of greenspace and trees (most of which were planted by residents in the 1970s) now dominating these supposedly "promethean" spaces, and largely protected from real estate developers through citizens' groups.[31] Since capitalism has flooded nominally communist Vietnam, however, the Vinh settlement, which lies at the heart of the city, is now coveted by venture capitalists, speculators and developers – by setting aside nature-infused open spaces that are not supposed to "produce" "value" other than allowing for a harmonious *Wechselwirkung* between human and non-human, the East German urban planners have instead created a literal bonanza which is soon to be carved up and sold off as the last available open space with greenery in Vinh, a city that has otherwise come to look like most other overcrowded developing world cities. Again, one part of the equation effects the other, and back again, in ways difficult to predict.

Another ecology emerges in the work of Michel Dupuy – an *information ecology*. A term first suggested by information and library scientist Bonnie Nardi,[32] the flow of information became as much a sensitive issue in East Germany as the environment. And in particular, the flow of information *about* the environment became, ultimately, explosive, according to Dupuy. Dupuy's work focuses on paradox that the GDR was actually quite open about research into industrial emissions and the damage they were causing to the environment. In part this was because the Party line had been that pollution was a legacy of capitalist industry, and many of the facilities the GDR had inherited dated to capitalist era, like the infamous Espenhain factory. So, the media published stories about citizens openly debating the problem of pollution and how it should be best tackled in

30 See Rubin, *Amnesiopolis: Modernity, Space, and Memory in East Germany* (New York: Oxford University Press, 2016).

31 See Juliane Richter, Tanja Scheffler, Hannah Sieben, eds., *Raster Beton: Vom Leben in Großwohnsiedlungen zwischen Kunst und Platte. Leipzig-Grünau im internationalen Vergleich* (Weimar: M Books, 2017).

32 Nardi, *Information Ecologies: Using Technology with Heart* (Cambridge, MA: MIT Press, 1999).

a socialist society. The rise of environmentalism as a powerful political force in the West, and one that was anti-growth and not friendly towards the GDR, suddenly made any discussion about environmental pollution a matter of national security, in the eyes of the Stasi.

As soon as censorship began, underground flows of information also began – like water, or air, information flows through the ecology as well, creating niches and systems. Dupuy writes about the Radio Glastnost station in West Berlin which broadcast the missing information about the ecology, smuggled out of East Germany to the West, back into the East. The ecological groups that developed did so not just out of concern about the environment, Dupuy makes clear – they did so because their government was lying and covering up the information that they could sense with their own bodies. Indeed, most of the acts of resistance of groups like Arche Nova were informational – it is not coincidental that perhaps the center of the underground resistance in East Germany was a basement *library* – the environmental library in Prenzlauer Berg, and that many, including Tobias Huff, have suggested the Stasi raid on the library as being the real beginning of the *Wende*.[33] Again, the *information ecology* mirrored the human-non human ecology, and both ecologies operated together within the broader context of the Cold War, in which what happened on one side had a profound impact on what happened on the other.

Moranda also explores a complex informational ecosystem, though in this case one that soon collapsed into informational monoculture. Marxist agrarian thinkers had long engaged in debate with agronomists, geographers, and economists of all political stripes. In fact, Marx, Kautsky, and others shared many of the pressing concerns about *Raubbau* that seemed to be stripping complex interconnected systems and replacing them with barren wastelands devoid of life. Georg Bela Pniower drew on those ideas in his proposals for small-scale agriculture in postwar Germany that would heal the metabolic rift created by capitalist agriculture.

Huff's work largely agrees with Dupuy's, in pointing to a much more complicated narrative of the GDR and its environment, noting that the

33 Huff, *Natur und Industrie im Sozialismus*, 399.

GDR was actually ahead of West Germany (or almost any other country) in attempting to address the problem of environmental pollution. The GDR began passing environmental legislation, and even created a Ministry of Environmental Protection and Water Management in 1973, more than a decade before the Federal Republic created an Environmental Protection Ministry and began passing serious environmental legislation. The problem, Huff notes, was not one of "prometheanism" or even necessarily "growth" but one of capital and technology – the advantage that West had was access to flows of capital which allowed them to develop and pay for technologies, such as scrubbing and de-sulfurization filters on smoke stacks, which great improved the air quality in the West, whereas East Germany did not have access to such technologies.

Rubin agrees, describing a "Green Legend" which, like the original "Black Legend," had elements of truth but took the pollution in the GDR out of context, using it to argue that the GDR, and socialism itself were promethean and therefore set up inherently to destroy the environment and eventually itself. Indeed, the consensus in public discourse in Germany (and beyond) was the *inverse* of Marxian notions of capitalism, socialism, and the ecology: it was *capitalism* that was a better steward of the environment because it could recognize the limits to resources since it recognized the value of those resources (indeed, the market was considered part of "nature" as well), whereas socialism's penchant for freebies was enabled by its prometheanism and the result was ecological, economic, and political collapse. Rubin's argument is that this "Green Legend" is in need of deconstruction, as are all dominant narratives, because it obscures more – especially now – than it illuminates. Constructing a genealogy that takes into account the complicated ecology of actors, states, humans and non-humans, and ideas is how Rubin approaches this task of unraveling the Green legend.

Thomas Fleischman concurs that there is a "greenwashing" of the FRG that has taken place since the *Wende* in Germany. As Fleischman writes, the GDR became a convenient punching bag for its environmental pollution, though many of the same problems existed in the FRG. Ecological metaphors proliferated in the discourse of the post-*Wende* years, that is, Kohl's discussion of "blossoming" in the former GDR. In reality,

Fleischman writes, the *Wende* represented a massive outflow of wealth and capital from the East to the West. As it did so, the factories of the former GDR closed, one by one, laying off millions of people, and also drastically cutting pollution – and in this way, capitalism took credit for cleaning up the environment sullied by pollution, when in reality, the flow of capital caused heavy industry to leave and relocate elsewhere – out of the places where Germans might have to experience its pollution.

Indeed, the former East has seen a reversion to relying on lignite, and the massive agribusiness that has developed in Germany puts any accusation of "gigantomania" in socialism to rest. Yet, perhaps most ironically, is the fact that it is the former East that is one of Germany's greenest areas, with massive greenbelts and nature preserves, many of which were written into law as the very last acts of the East German legislature before dissolving itself. In fact, these massive protected areas were the result of the same green activists, including many devotees of the Kretschmanns, who participated in the *Wende* and the Roundtable, and persuaded the disintegrating *Volkskammer* to preserve space from future real estate development – precisely what never happened in Vinh, because communism never "collapsed" so there was never a clear *Wende*. Even more ironically, it is the preservation of so much unbroken green space that has allowed the migration of massive amounts of wild boars who are, it turns out, radioactive, with the fallout from Chernobyl – which was the most recognizable symbol of the dangers of Soviet prometheanism as well as validation of the origins of the Green Party in the anti-nuclear protests of the 1970s.

Gernot Waldner begins his essay discussing the foundation of the Green Party in Austria. The story was similar to that in Germany: an occupation to stop a power company building a new plant which would destroy a natural wetlands area, following by the founding of a political party by many of the same activists. Like the West German Greens, the Austrian Greens were not a "left" party – in part, because their actions were directed at the pro-growth policies of the Social Democratic regime that had been in power, and which was supporting expansion of industrial and technological facilities. In part, because there were rightists involved in the protest and the founding of the Party, including those with a Nazi past. Waldner describes this as an "ecological assemblage" without a "coherent ideology" and which was more like a "feeling." This ecology of ecology would lead, he writes, to the "political wilderness," at least for the Left.

Waldner's focus is the writer Elfriede Jelinek, a member of the Communist Party in Austria, who watched the occupation of the threatened Hainberger Au wetlands and the founding of the Greens, and subsequently wrote *Oh Wildnis, Oh Schutz vor Ihr!* [Oh Wilderness, Oh Protect us from It!], a novel written in Jelinek's experimental, nonlinear style. *Oh Wildnis* is a reflection on the complexities of the role that the ecological movement played in Austrian politics, but also in Austrian political economy and society. It is not a straightforward Marxist critique of anti-growth Greens, as we might expect from a Party ideologue. It is not straightforward anything – much like the ecology itself, and the politics of ecological movements. It is both about "nature" *and* about how discourses of nature shape language – and it is written in an "ecological" style precisely to disrupt those discourses, as Waldner describes her style: "The form of the text emulates a wilderness, in spite of the content being concerned with nature. The entire text is written in this impertinent *genus humile*, assembling proverbial malapropisms, rural idioms, and pithy sarcasm. These choppy bits and pieces of language are the means as well as the target of Jelinek's political intervention ..." The book, as many critics complained, has no plot – like nature, it has no end or beginning. Jelinek does not really even want readers to "think" so much as she wants them to react to the toxic mix of politics and nature, with their bodies – she wanted her readers to "vomit" from reading her work, as Waldner writes.

Indeed, in a way, Jelinek's attempt to understand the bizarre tangle of ideas and forces that led to the ecological movement in Austria within the bigger ecology of ecology itself, leads her to collapse metaphor into reality, thereby bringing her close to Jason Moore's ideas about work/energy in both human and non-human nature. She writes about the disappearance of the working class in real nature – as bodies buried beneath the soil, and in ecological discourses that displace the exploitation of people for a concern about the exploitation of trees, when in reality, the trees feed off the bodies of the masses who were exploited; in other words, both trees and proletarians are workers, whose energy is exploited for the comfort and prestige of the bourgeoisie. Though Jelinek and her work are centered in Austria, not Germany, Waldner's essay shows how the core themes that this volume explores extend into Austrian society and politics – the political ambiguity of "green" movements and their potential to confound and split the traditional Left by mobilizing older, romantic notions of nature, is not a phenomenon that is limited just to West Germany's unique Cold War position. It extends, geographically as well as diachronically.

Katrina Nousek's essay also considers the way that the complexities of the ecology, Green politics, socialism and post-socialism, and capitalism, form a web or an ecology themselves, and again, through literature – in this case, Ingo Schulze's *Simple Storys*, a collection of loosely connected vignettes set in post-*Wende* East Germany. Nousek directly invokes Moore, seeing Schulze's characters as existing in a "web of life" which has no center, no main themes or main characters, like Jelinek's work, and one in which non-human actors repeatedly make their presence known. Not only is reality in post-*Wende* Germany explored as a web with no center, tying together humans and non-humans, but it is an *absurd* web, often tragically so. One character runs over a badger with her car, but ends up have to imitate a badger to describe the incident, and in so doing, her increasingly absurd bodily movements make it clear she does not know what a badger looks like, and the darker truth is hinted at: she actually ran over a person. Is it more acceptable to talk about how we hurt nature than how we hurt other humans in postsocialist East Germany?

Indeed, Nousek directly relates the characters in *Simple Storys* to Moore's notions of "environment-making" as they try to adapt to a radically new environment of relationality in the postsocialist world, like flora or fauna taken out of their niches, out of their *oikeios topos*, to harken back to Theophrastus. Such re-orientation to a new web of relationality is, in Schulze's work, by necessity produces absurdity. Or rather, it reveals the inherent absurdity of neoliberal dualism – the idea that humans are individuals, in control of their environments, moving in a linear way, making their environments, but not made by their environments.

Finally, cultural anthropologist Bettina Stoetzer takes us into the ecology of post-socialist landscapes, specifically, the "wilds" of Brandenburg, the state that surrounds Berlin and forms its margins and hinterland. There she meets Carsten, the "Computer Indian," who grew up in the GDR and found his dreams of self-actualization difficult to attain after unification. Becoming a computer programmer, he was not sad to see the GDR go, but found capitalism just as restrictive. So, he has carved out a bittersweet existence as an "Indian," attempting to find his own postcapitalist and postsocialist existence through a rejection of modernity, consumer culture, and modern work culture, living in harmony with Brandenburg's forests as he imagines Native Americans lived in the forests of North America. The irony is not lost on Stoetzer, of course: the "colonization of the GDR" by the West is now a common frame of reference for former East Germans, the fascination with the "wild West" was itself a frame of reference for German colonialism, in socialist East German westerns, the Indians were the heroes and the white settlers were the bad guys, and much of the change that has swept over rural, "wild" Brandenburg has the feel of a colonization by urbanites from Berlin.

Highlighting the poignant absurdity of Carsten's story is the nearby refugee center, built in a former East German army base in what East African asylum seekers call the "bush" in Brandenburg, where they and other people from Africa, Asia and the Middle East are forced to live in isolation, for years while their cases are adjudicated. There, these refugees – many of whom are from post-colonial spaces, fleeing the instability and deprivation resulting ultimately still from colonialism and its legacies, find themselves in spaces defined by the neoliberal state. Here, in the margins and the wilds, Stoetzer shows us the absurdity and tragedy at the heart of this state, which readers might even associate with a new form of colonialism.

The essays in this volume form a kind of ecology: sometimes, they flow together, and sometimes they clash. In this clashing and flowing, they reenact the dynamics that characterize the interactions of "red" and "green" thought and movements in German history and culture, and beyond. Within their "ecosystem," they consider the problem of socialism and ecological thought through different disciplines and across multiple historical and cultural contexts. The result, hopefully, is an opening onto the ecology of socialisms in Germany.

SCOTT MORANDA

A Garden of Small Plots or Factory Farms? Early Cold War Agricultural Planning in East Germany

Agricultural modernization radically transformed rural landscapes on both sides of the Iron Curtain, as agronomists and farmers embraced tractors, chemicals, and larger farms all across postwar Europe. Even if small family farms were more likely to survive under capitalism, western government policies and market forces still led to the consolidation of small plots into ever larger farms. Biodiversity plummeted on the resulting farms, as a mix of pastures, cropland, hedgerows, small streams, orchards, and woodlands gave way to monoculture and specialization. Synthetic fertilizers led to the nitrification of streams and ponds to the detriment of fish and other wildlife.[1]

Although these trends were global, the ills of industrial agriculture more often feature in the histories of socialism and its failures. In part, this reflects real differences between East and West. Under centralized planning and with a socialist admiration for large-scale projects, state planners in East Germany pushed through immense changes in a very short time-period. High modernist planning reoriented farms around machines and transformed rural

1 See Frank Uekötter, "Why Panaceas Work: Recasting Science, Knowledge, and Fertilizer Interests in German Agriculture," *Agricultural History* 88, no. 1 (2014): 68–86; Arnd Bauerkämper, "The Industrialization of Agriculture and its Consequences for the Natural Environment: An Inter-German Comparative Perspective," *Historical Social Research* 29, no. 3 (July 2004): 124–149; Mark Finlay, "New Sources, New Theses, and New Organizations in the New Germany: Recent Research on the History of German Agriculture," *Agricultural History* 75, no. 3 (2001): 279–307; Kiran Klaus Patel, *Europäisierung wider Willen: Die Bundesrepublik Deutschland in der Agrarintegration der EWG 1955–1973* (Munich: Oldenbourg, 2009); Ulrich Kluge, *Vierzig Jahre Agrarpolitik in der Bundesrepublik Deutschland* (Hamburg: P. Parey, 1989).

spaces through collectivization.[2] Later, agricultural authorities increased meat production through large-scale confined livestock farming.[3] Indeed, the SED's Ministry for Agriculture saw nature preservationists and traditional farming as obstacles to the modernization they believed was required in order to provide more food to workers and to enlighten the countryside.[4] Given the speed and scale of these changes in East Germany and the Soviet Union, journalists and historians often link industrial farming to the collapse of socialist economies and ecosystems.[5] Less often, historians highlight similarities in factory farming on both sides of the Iron Curtain.[6]

2 See Arvid Nelson, *Cold War Ecology: Forests, Farms, and People in the East German Landscape, 1945–1989* (New Haven, CT: Yale University Press, 2005); George Last, *After the 'Socialist Spring': Collectivisation and Economic Transformation in the GDR* (New York: Berghahn Books, 2009); Andreas Dix, *Freies Land: Siedlungsplanung im ländlichen Raum der SBZ und frühen DDR 1945–1955* (Cologne: Böhlau, 2002).

3 See Patrice G. Poutrus, *Die Erfindung des Goldbroilers: Über den Zusammenhang zwischen Herrschaftssicherung und Konsumentwicklung in der DDR* (Cologne: Böhlau, 2002); Thomas J. Fleischman, "Three Little Pigs: Development, Pollution, and the „Greening" of East Germany, 1970–1989" (PhD dissertation, New York University, 2013), Dissertation Abstracts International.

4 See Scott Moranda, *People's Own Landscape: Nature, Tourism, and Dictatorship in East Germany* (Ann Arbor: The University of Michigan Press, 2014); Gerhard Krenz, *Notizen zur Landwirtschaftsentwicklung in den Jahren 1945–1990: Erinnerungen und Bekenntnisse eines Zeitzeugen aus dem Bezirk Neubrandenburg* (Schwerin: Ministerium für Landwirtschaft und Naturschutz des Landes Mecklenburg-Vorpommern, 1996).

5 See Murray Fesbach and Alfred Friendly, Jr., *Ecocide in the USSR* (New York: Basic Books, 1993); "Ecocide in the USSR," Foundation for Economic Education, accessed June 1, 2018, <https://fee.org/articles/ecocide-in-the-ussr/>; Roger Manser, *Failed Transitions: The Eastern European Economy and Environment Since the Fall of Communism* (New York: New Press, 1993); "Capitalism, Communism, and Environmental Protection: Lessons from the German Experience" *Environmental History* 3, no. 3 (1998): 311–332; Joan DeBardeleben, *The Environment and Marxism-Leninism: The Soviet and East German Experience* (Boulder, CO: Westview Press, 1985); Charles Maier, *Dissolution: The Crisis of Communism and the End of East Germany* (Princeton, NJ: Princeton University Press, 1997), 172.

6 One example of an article focused on similarities is: Kate Brown, "Gridded Lives: Why Kazakhstan and Montana are Nearly the Same Place," *The American Historical Review*, 106, no. 1 (February 2001): 17–48.

With their focus on high modernism, historians have done less to explore alternatives to collectivized agriculture that were either implemented, or considered for implementation, in socialist countries. The failures of collectivization in Poland and Yugoslavia mean that histories of agriculture in those countries do focus on small farms, but collectivized agriculture had its limits even in other socialist states.[7] Diana Mincyte, for example, has written about the importance of gardens and small operators to the food supply in Lithuania.[8] Thomas Fleischman demonstrates the importance of gardens and small operators for East German food production.[9] These histories of surviving small-scale operations (and their supporters) help us more fully understand the ecological consequences of socialist agricultural policy.

One prominent East German author actually published work on small farms and gardens in 1948. In *Bodenreform und Gartenbau*, Georg Bela Pniower proposed creating urban and rural landscapes full of small farms and gardens.[10] He was not alone. Within the scientific institutes associated with the GDR's Ministry of Agriculture, biologists and landscape planners in the 1950s continued to praise the virtues of a mixed landscapes with small plots bordered by hedges and forests, often called a "*Heckenlandschaft*." Landscape architect Walter Funcke, for example,

7 See M. C. Maurel, "From Disappearance to Recovery: Family Farming in Central Europe. Questioning the Path Dependency Hypothesis," *Village and Agriculture (Wieś i Rolnictwo)* 166 no. 1 (2015): 79–105; Krzysztof Gorlach, Piotr Nowak, Martyna Wierzba-Kubat, "The Art of Survival: Family Farms in Poland," *Agriregionieuropa*, 11, No. 43 (December 2015); D. Turnock, "Agriculture in Eastern Europe: Communism, the Transition and the Future," *GeoJournal* (1996) 38: 137–149; J. B. Allcock, "The Socialist Transformation of the Village: Yugoslav Agricultural Policy Since 1945," in *Agricultural Policies in the USSR and Eastern Europe*, ed. R. A. Francisco, et al. (Boulder, CO: Westview, 1980), 199–216.

8 Diana Mincyte, "Everyday Environmentalism: The Practice, Politics, and Nature of Subsidiary Farming in Stalin's Lithuania," *Slavic Review*, 68, no. 1 (Spring 2009): 31–49.

9 Fleischman, "Three Little Pigs."

10 Georg Pniower, *Bodenreform und Gartenbau* (Berlin: Siebeneicher Verlag, 1948), 33, 63, 71.

celebrated mixed agricultural landscapes for the preservation of soil fertility.[11] Pniower was unique, however, in that he had a commitment to Marxism even before 1945.

Rather than an in-depth analysis of Pniower's career and writings, this chapter uses Pniower's 1948 book as a jumping off point for reconsidering how we periodize and contextualize socialist agricultural thought. First, it shines a light on a book that was largely forgotten as collectivization intensified. In certain respects the book was anomaly, but it also built on the work of earlier socialist agronomists. By exploring similarities and differences with earlier socialist agrarian writing, this chapter makes a case for the diversity of early Cold War agricultural thought. Second, the chapter will link socialist agricultural thought to a wider debate about the causes of soil decline. In the late nineteenth and early twentieth centuries, economic liberals, social conservatives, and Marxist socialists all closely studied American surpluses of wheat and other commodities to understand their implications for Europe. Socialists, in fact, began to write about social metabolism as a result of their participation in this larger conversation about American national character, colonial frontiers, ethnic difference, land ethics, and soil fertility.[12] If Pniower's musings on small-scale agriculture departed from Marxist orthodoxy in certain ways, the book's perspective was fully shaped by a Marxist emphasis on soil plunder and the so-called metabolic rift. Third, the chapter argues that, while Marxist land-use policy after 1950 reflected certain ideological constants in Marxist thought, it cannot be fully explained by ideology. Pniower's book was a product of a distinct period from 1945 to 1950 that was characterized by physical destruction, material scarcity, and efforts to reimagine German culture and politics. Such conditions often led to creative criticism of key assumptions about modernization and growth so central to the post-Enlightenment worldview. As the Cold War hegemons pivoted to

11 See Susanne Karn, *Freiflächen- und Landschaftsplanung in der DDR: Am Beispiel von Werken des Landschaftsarchitekten Walter Funcke, 1907–87* (Münster: LIT Verlag, 2004).

12 On socialist commentary about metabolism and soil mining in North America, see John Bellamy Foster, *Marx's Ecology: Materialism and Nature* (New York: Monthly Review Press, 2000), 152, 236–240.

outproduce their ideological enemies, such creativity declined. Marxism's prometheanism was not inevitable (even if it was always possible). The unique circumstances of postwar western Europe mattered as much or more than ideological rigidity for the resulting agricultural policy.

Pniower's Contributions to East German Landscape Architecture

Georg Bela Pniower is largely forgotten today, but his work profoundly shaped the landscapes of the early GDR. As he had been categorized as a "half Jew" by the Nazis, the Third Reich banned him from professional organizations and sentenced him in 1944 to slave labor in a textile factory. Being a Social Democrat, Pniower's Marxism worked more in his favor after the war. While he initially worked in an institute in Berlin-Dahlem in West Berlin, he relocated in 1951 to Humboldt University in the East to head a department of Garden and Landscape Architecture. Between 1945 and 1960, he was one of the most prominent landscape planners helping to rebuild Berlin and the rest of the Soviet-occupied zone, working alongside Frank Erich Carl and Reinhold Lingner, two other landscape architects with roots in the garden planning and landscape care of 1920s Weimar Germany. With these associates, Pniower planned Berlin's "rubble mountains," developed early concepts for rehabilitating old strip mines in East Germany's coal district, helped build youth summer camps, and engaged in the restructuring of rural agricultural landscapes. In addition to occupying the GDR's most prominent academic position in landscape architecture, Pniower made a name for himself in debates over the environmental protection law of 1954 and in his work with the agricultural ministry in the development of model collective farms.[13]

13 For an overview of Pniower's career, see Hermann Behrens, "Landeskultur als Naturgeschehen auf höherer Ebene: Georg Bela Pniower (1896–1960) und der

Pniower's work reminds us of two important aspects of East German agricultural and environmental history. First, advocates of industrialized agriculture and collectivization did not reject nature protection completely. Despite the often nasty battles to discredit "reactionary" colleagues, both sides of that struggle often shared concerns about environmental degradation. Second, East German approaches to landscape planning were not static. While Pniower later enthusiastically endorsed Stalinist collectivization, he followed a winding path to get there.

In *Bodenreform und Gartenbau*, Pniower sharply criticized the concentration of capitalist industry in large cities. He described the growth of dense urban centers where workers often lived close to the smokestacks belching out toxic fumes or drank from contaminated streams while wealthier landowners escaped to greener neighborhoods. Growing cities and suburbs gobbled up former agricultural land, so that agriculture shifted further and further away from the urban core and away from the majority of the population. As a result, fertile soil near industrial cities was wasted, and distant landscapes experienced increasingly intense plunder to feed hungry workers and middle-class desires. Pniower linked military aggression and authoritarian politics to this aspect of the country's unsustainable economy. Unable to provide resources to cities growing uncontrollably, Germany mobilized for war and territorial conquest to make up for those shortages. According to Pniower, Germany needed to consume less overall and produce as much as its own food as possible in order to truly democratize and end imperialism.[14]

As an alternative, Pniower proposed a small-scale, decentralized economy. He strongly criticized reconstruction plans that attempted a restoration of prewar conditions, such as proposals to increase the export of manufactured goods from rebuilt industrial centers in order to pay for imports of calories. Instead, he imagined that planners would no longer separate farms, factories, and homes from each other. Housing and industry would coexist with green spaces and continued use of fertile soils once

Naturschutz" in *Naturschutz hat Geschichte*, ed. Bernd Kreuter (Essen: Klartext Verlag, 2003).

14 Pniower, *Bodenreform und Gartenbau*, 34.

threatened by urban sprawl. Politically, such a rearrangement of the landscape would bring farmers and workers closer together and eliminate the distrust between the two classes. Moreover, it would improve the living conditions of the working class.[15]

Intense "gardening" within a semi-urban/semi-rural landscape, he believed, also promised the most sustainable and productive agricultural system for Germany. He described how small farms better cared for the land. Mixed use farms growing vegetables and grains alongside small herds of dairy cattle maximized productivity and soil fertility. As Pniower argued, a transition away from large-scale grain farming to small-scale mixed farms would benefit the soil, as small farms and gardens better tended the soil with their careful use of compost, manure, crop rotation, and other soil improvement measures.[16] In addition, small farms and gardens could be better integrated with light industry and housing, thus providing beauty, clean air, and more elbow-room to the working class.[17]

Pniower was not the only landscape architect in East Germany with kind words for small farms. Walter Funcke, for example, worked to return rural landscapes to a healthy balance between nature and society. He produced plans for a model village situated in a landscape of hedges and mixed-use small farms and argued that a landscape of intensive 5- to 10-hectare "garden" farms would best solve the GDR's food supply concerns. Large mechanized farms, he argued, had never taken natural conditions and landscape health into consideration.[18]

Given Pniower's arguments in *Bodenreform und Gartenbau*, one would think he would have allied with Funcke and other like-minded planners. Instead, Pniower in the 1950s worked furiously to discredit landscape architects and nature preservationists such as Funcke. Even as Pniower promoted nature protection, he reported to the Ministry of Agriculture about the problematic views of other preservationists. While many architects before and after World War II worked to cure "sick" landscapes and recreate a

15 Pniower, *Bodenreform und Gartenbau*, 33, 71.
16 Pniower, *Bodenreform und Gartenbau*, 56–57.
17 Pniower, *Bodenreform und Gartenbau*, 76.
18 Susanne Karn, *Freiflächen- und Landschaftsplanung*, 216.

"healthy" diversified landscape to improve public health, Pniower condemned "the application of a medical term such as *diagnosis* in the study of a geographical milieu." He added, "This term can give the impression that the landscape … is an 'organism' that can get sick. This is not the case."[19] Such language, Pniower elsewhere suggested, could only mean that landscape architects had not overcome their racist past, their associations with imperialism, or their affinity for a "blood and soil" ideology that placed healthy Aryan bodies within "healthy" landscapes and linked racial inferiors to "sick" landscapes.[20] Pniower also insisted that unreformed landscape architects did not do enough to eliminate capitalist land tenure in the countryside. In the late 1950s, he led a project at Humboldt University to improve the eroded landscape of the Huy Hakel region on the edge of the Harz Mountains. In this project, Pniower clearly celebrated industrialized collective agriculture and scientific expertise for overcoming the ignorance of the poor capitalist farmer to eliminate erosion and flooding.[21]

19 Pniower, quoted in Axel Zutz, "'Kranke' und 'gesunde' Landschaft – Anmerkungen zur Kritik des Landschaftsbegriffs bei der Landschaftsdiagnose," in *Die Landschaftsdiagnose der DDR: Zeitgeschichte und Wirkung eines Forschungsprojekts aus der Gründungsphase der DDR*, ed. Olaf Hiller (Berlin: Technische Universität Berlin, 2002), 112; Zutz takes this from the "Protokoll über die Arbeitstagung der Fachrichtung Garten und Landschaftsgestaltung am 5. und 6.1.1957 aus Anlass des 10jährigen Jubiläums der Fachsschule Erfurt," January 6, 1957, Nachlass Rindt. See also Bundesarchiv (BArch): DK107 25/13, Georg Pniower, "Bemerkungen zur Problematik der Landeskultur," March 1956, 27. For more background on Pniower, see Behrens, "Landeskultur" as well as Klaus-Dietrich Gandert, "Georg Bela Pniower – sein Leben und Wirken für die Garten- und Landeskultur," in *Landschaft und Planung in den neuen Bundesländern – Rückblicke*, ed. Institut für Umweltgeschichte und Regionalentwicklung e. V. (Berlin, Verlag für Wissenschaft und Forschung, 1999), 221–235.

20 Moranda, *Peoples' Own Landscape*, 49–78. Indeed, landscape planners in both East and West Germany continued to employ such vocabulary and concepts after the war, though advocates of healthy landscapes varied in their ideologies. Many even had clashed with the Third Reich at key moments. On the relationship of landscape architects to Nazism, see Thomas Lekan, *Imagining the Nation in Nature: Landscape Preservation and German identity, 1885–1945* (Cambridge, MA: Harvard University Press, 2003), 204–251.

21 BArch DK107 25/13, Georg Pniower, "Bemerkungen zur Problematik der Landeskultur," March 1956, 8.

Given Pniower's ideological attacks, historian Susanne Karn presents him as a foil to Funcke, but there are elements of Pniower's career that suggest a different narrative that makes more room for ambiguity and contingency. Pniower, Karn writes, only appreciated intensely managed small farms and gardens as a short-term strategy that would eventually give way to "pure monocultures" of vegetables and other crops managed scientifically with the help of cutting-edge technology.[22] Given Pniower's attacks on colleagues such as Funcke for their association with Nazi planning, Karn's depiction makes a lot of sense. Pniower did help to silence criticisms of collectivization and modernization if he sensed that those critics clung to worldviews tainted by Nazism. Yet, Karn downplays similarities between the two planners. Not only did Funcke own his own copy of *Bodenreform und Gartenbau*, but Pniower' extensive praise for small farms in that very book also resembled Funcke's ideas to a remarkable degree. Other appraisals of Pniower's career make more room for ambiguity. Unlike Karn, historian Andreas Dix stresses Pniower's concern for natural environments. Dix, in fact, emphasizes Pniower's commitment to nature protection as revealed in a 1948 speech, in which Pniower explained how an intensive farming in a mixed landscape of small farms, hedges, and forests was better adapted to local conditions and better protected soil from erosion.[23]

Plunder and American Agriculture

Were Pniower's ideological battles evidence of socialism's promethean disdain for the natural world, or does his story reveal that the trajectory of socialist agriculture remained uncertain in the 1940s? His attacks on other

22 Pniower, *Bodenreform und Gartenbau*, 76.

23 Andreas Dix, "Nach dem Ende der 'Tausend Jahre': Landschaftsplanung in der Sowjetischen Besatzungszone und frühen DDR," in *Naturschutz und Nationalsozialismus*, eds. Joachim Radkau and Frank Uekoetter (Frankfurt/Main: Campus, 2003), 352.

landscape architects might suggest that Marxists such as Pniower dismissed all nature protection out of hand. Yet, other publications and comments by Pniower suggest otherwise. When situated within the longer history of socialist agricultural thought, the complexity of socialist agricultural thought becomes even more apparent. Contextualizing Pniower's work, in fact, reveals that a socialist agricultural thinker attacking conservative agronomists and landscape architects could also, at the same time, recognize the value of small-scale farming systems that avoided plundering the soil's fertility. In particular, Marxist concerns about the "metabolic rift" deepened by capitalist plantations suggest that socialism could recognize limits to humanity's power to exploit nature without consequence. Marxist agronomists sometimes shared a respect for the careful soil care of the small farmers that were so often praised by non-Marxist conservationists. Rather than an insignificant blip outside the mainstream of GDR agricultural policy or an insincere short-term strategy Pniower's work in the 1940s becomes a link in a longer chain of Marxists grappling with the complexities of rural economies while remaining sensitive to biophysical limitations to human adaptation.

For understanding Pniower's debts to earlier Marxist writings and the potential for agreement with individuals such as Funcke, the idea of soil plunder is key. In *Bodenreform und Gartenbau*, Pniower comments on the plunder of soil by large landed estates dominated by aristocrats and capitalists. Here, as will be seen, Pniower draws on language that goes back to Marx himself. When Funcke used similar arguments to challenge monoculture landscapes advocated by the East German agricultural ministry, he was not just making a tactical ploy to win sympathy from his supervisors. He drew on a vocabulary he shared with Pniower and other socialist agronomists. While Marxist thinkers typically believed the small family farm was doomed, the ideological roots of Marxist and non-Marxist agricultural policy were much more entangled than is often appreciated.

German agronomists had been discussing plunder and soil mining for a long time, in fact. Beginning in the 1850s, Marxists and non-Marxists contemplated the remarkable transformation of the American economy and the sudden glut of American wheat in European markets. The matter of small farms stood at the center of the debate about American farming and its consequences for Europe. Plunder, or *Raubbau*, became a buzzword among travelers and writers across the political spectrum, as concerns about the waste of valuable natural resources grew.

Conservative geographers were among the first to ponder the colonial plunder of nature by western societies. Before and during Marx's lifetime, non-Marxist geographers contemplated modern commercial economies and their abuse of natural resources, with particular attention to colonial plunder in the Anglo-American world. Friedrich Ratzel's multivolume travel narrative about North America, for example, identified plunder as an essential characteristic of an "immature" American economy. Ratzel suggested that America would eventually evolve into a more mature economy, in part due to cultural diffusion. In a few brief passages, he also suggested that German-speaking immigrants offered a model for small-scale, permanent agriculture that would eventually save the United States from Anglo-Americans who devastated the natural abundance of North America through plunder.[24]

Under the influence of Ratzel, geographer Ernst Friedrich further analyzed the relationship between colonialism and plunder in 1904. Environmental historians have recently rediscovered Friedrich's prescient recognition that the irresponsible devastation of natural resources reverberated throughout an interconnected biophysical system. Even more than nomadism or war in his analysis, colonial expansion unleashed an orgy of environmental destruction. On colonial frontiers, settlers reverted to a "natural state" and abandoned rational farming for superficial soil mining. They "shipped" soil nutrients downriver and overseas as they produced commodities for export without any concern for future generations of farmers.[25]

As Ratzel did, Friedrich remained optimistic. Once soil began to decline, settlers adjusted and returned to labor-intensive practices to preserve fertility. In the United States, for example, the plunder economy persisted until population density increased and farmers ceased migrating

24 Friedrich Ratzel, *Die Vereinigten Staaten von Nord-Amerika* (Munich: R. Oldenbourg, 1878), 167, 245, 249.

25 Ernst Friedrich, "Wesen und geographische Verbreitung der Raubwirtschaft," *Petermanns Mitteilungen* 50 (1904), pp. 68–79, 92–5.

westward to empty lands. Repeatedly, Friedrich also suggests that such waste and destruction was the price to be paid for cultural evolution.[26] Friedrich's analysis, like many at the time, relied on racist assumptions about global cultures. Energetic Europeans, even if they first wasted resources on newly settled lands, eventually introduced rational, progressive land use once "less cultured" peoples were removed from the land.[27]

Elsewhere, Max Sering drew on his own observations of North American agriculture to praise small-scale agriculture and advocate for inner colonization in Germany. Sering, a professor of economics from Berlin University, argued that the distinct advantage of Canada and the United States was their ability to "revive" its eastern population through "inner colonization" of the West. He contrasted entrepreneurial North American farmers to German farmers, who he believed lacked creativity, industriousness, or innovativeness after centuries of oppression and exploitation. As Ratzel and Friedrich did, Sering discussed soil mining and the plunder of North American soils. As Friedrich did, he emphasized that plunder would decline as more efficient farmers took over bankrupt farms and once fertile, "untouched" soils were fully mined. In other words, the ability of North America to flood European markets with cheap wheat obtained through soil mining was soon to be over. As long as cheap land and plunder continued in North America, however, Germany needed to focus on playing to its strengths – small plots, careful farming, and high-value goods difficult to transport cheaply over long distances. Sering encouraged specialization in high-value products on small farms near urban markets and loudly attacked efforts to compete with American plunder on vast Junker estates worked by migrant labor.[28]

Sering's work, in addition, points to an additional characteristic of the plunder critique that became increasingly prominent among non-Marxists. He suggested in his conclusion that Germans had a particular advantage in their superior ability to carefully tend soil and properly manure fields. Often,

26 Friedrich, 72–73.

27 Friedrich, 95.

28 Max Sering, *Die landwirtschaftliche Konkurrenz Nordamerikas* (Leipzig, 1887), 181–188.

he also noted, German settlers (now freed of their European chains) led the way in the transition to better stewardship of the soil in the United States. Germany, he also argued, could learn little from America when it came to reforming land use; the worst thing German agronomists could do was to mimic American practices of extensive and careless farming. Given that he found German peasants wanting in their self-confidence, intelligence, and ambition, he must have believed that an "inner colonization" that would allow the best traits of German peasants (careful farming on limited land) to merge with the better characteristics of American "entrepreneurial" farmers.[29]

Increasingly, non-Marxist investigations of the "plunder economy" identified ethnicity as a crucial factor for overcoming waste. Going further than Ratzel did, many studies of American agriculture paid particular attention to the diffusion of ethnic German agricultural practices. The Austrian conservative journalist and "Catholic socialist" Rudolf Meyer largely shared the optimism of Friedrich and Ratzel, but he paid more attention to ethnicity and nationality than they did. In his journalistic and scholarly reports on American agriculture, Meyer hoped to rebuff free trade advocates who suggested that American competition was not really a long-term threat to Europe since American soil was quickly losing its potency. In response, Meyer (as Sering did) explained how America had already begun to abandon a plunder economy.[30] More emphatically than Sering, Meyer argued that new immigrants from Europe sparked this transition as they settled near careless American farmers.[31] While Anglo-Americans had been reckless speculators who wasted resources on consumer comforts, German farmers carefully tended the soil and avoided debt, producing yields two to three times larger than their Anglo-American neighbors. When poorly managed American farms went under, Germans swooped in to purchase the land.[32]

29 Sering, 710–715.

30 Rudolf Meyer, *Ursachen der amerikanischen Konkurrenz: Ergebnisse einer Studienreise durch die Vereinigten Staaten* (Berlin: Bahr, 1883), 37, 83, 118.

31 Meyer, 118.

32 Meyer, as Sering did, focused on the productivity of these German-American farms due to his hopes for homesteading in Europe. He praised these German homesteaders who recovered the independence and self-reliance that had been stolen from them

For German-American agronomists and community leaders writing in German-language newspapers and pamphlets published in Milwaukee and other heavily German cities in the Midwest, their own praise of a German land ethic also addressed a pressing existential concern. Facing the constant threat of Americanization, they believed that a careful tending of the soil was essential for keeping tight-knit German-speaking communities from breaking apart. Linguistic islands of permanent settlement only survived if German-speakers passed on prosperous farms to the next generation and provided incentives for young people to remain rooted in the community. As a result of these fears of cultural dissolution, German-Americans were especially keen to link ethnicity to land use practices. Not only did this contribute to ethnic pride, but contrasting German-American and Anglo-American land use practices might help enlighten the worst German farmers. If all Germans farmed better, ethnic islands could be preserved and even expanded at the expense of local non-German farms.[33]

Karl Marx engaged in a robust debate with his fellow observers of American waste and plunder. He corresponded with Rudolf Meyer, for example. He also passionately read the work of Justus von Liebig and other commentators on the American agricultural scene. From his daughter, he sought to learn as much as he could about the American scene after her speaking tour of the United States. As Andrew Zimmerman has demonstrated, many German radicals relocated to the United States, where they continued to develop their revolutionary theories about capital, land, and the use of violence to enact social change. Many of

---

by 300 years of bureaucratic paternalism in the "old country." He hoped to replicate the "homesteading" idea in Europe. Meyer, 356–66, 571.

33 This literature is immense. You can find these comparisons in German-American agricultural newspapers such as *Haus und Bauernfreund* published in Milwaukee. In addition, several pamphlets and books written by German-Americans repeated these comparisons. One example was Ernest Bruncken, *Germans in America* (Washington: G. P. O., 1899). One introduction to European immigrants and their critique of American agriculture can be found in Jon Gjerde, *The Minds of the West: Ethnocultural Evolution in the Rural Middle West, 1830–1917* (Chapel Hill: University of North Carolina Press, 1997).

these individuals remained in correspondence with their colleagues in central Europe.[34]

Marx's observations of American agriculture became central to his understanding of capitalism. At first Marx generally agreed with Ratzel and Meyer, in that he believed that education, scientific progress, and new technology could continuously improve American farming practices. The plunder of North American soils, for the younger Marx, could be explained by the immaturity of the economy or the under population of the vast North American continent. It would eventually end.[35]

Marx, as John Bellamy Foster and Kahei Seito have demonstrated, revisited the issue in the 1860s. Justus von Liebig had profoundly influenced Marx's earlier optimism through his suggestion that synthetic fertilizers could overcome soil decline. But, in his later edition of *Agricultural Chemistry*, Liebig strongly criticized the continued plunder of soils in industrial economies. Following Liebig, Marx also abandoned his optimism. He recognized that new technology and scientific advancements did not necessarily put an end to soil exhaustion. Instead, he concluded, the capitalist system recklessly perpetuated plunder. It was built into the system. Revisiting another important source, Johnston's *Notes on North America*, Marx in the 1860s paid more attention to Johnston's descriptions of land exhaustion, particularly to farmers who prioritized profits over future fertility to provide goods for distant markets. Since farmers shipped their products long distances, they had no opportunity to recycle nutrients drawn from the soil used to grow those crops. The nitrogen and phosphorous drawn out of the soil to grow wheat in Iowa traveled by way of grain exports to the hungry mouths of English workers and finally

34 See Andrew Zimmerman, "From the Rhine to the Mississippi: Property, Democracy, and Socialism in the American Civil War," *Journal of the Civil War Era*, 5 Issue 1 (March 2015): 3–37; Rachel Holmes, *Eleanor Marx: A Life* (London: Bloomsbury Press, 2015).

35 See Kohei Saito, "The Emergence of Marx's Critique of Modern Agriculture: Ecological Insights from His Excerpt Notebooks," *Monthly Review* 66 (October 2014). Correspondence with and about Meyer can be found in Friedrich Engels and Karl Marx, *Karl Marx, Frederick Engels: Collected Works* (New York: International Publishers, 1975).

ended up being pumped into the Thames in London. Unlike Johnston, however, Marx refused to believe that plunder was a short-term and necessary stage of development. Instead, he feared it was a permanent feature of capitalist agriculture.[36]

Marx's thinking about land and (what we would call) sustainability thus evolved over time. Marx still believed that highly developed, rational cultivation of the land could allow societies to prosper without robbing the soil of nutrients. He still assumed that cultural innovation and technological discoveries could allow for continual growth. Maluthusian fears of resource exhaustion were, in other words, still overblown and reactionary from Marx's perspective. Nonetheless, Marx's later writings gave much more attention to the notion of a metabolic rift. Both workers and the soil were sacrificed for profits. Marx wrote, "The same blind eagerness for plunder that in the one case exhausted the soil, had, in the other, torn up by the roots the living force of the nation [the working class]."[37] As long as farmers provided commodities to great capitalist population centers, Marx believed that a rift in natural metabolism would continue. The natural recycling of nutrients proved impossible when individual farmers seeking profit shipped goods from the American Midwest or the Irish countryside to distant urban markets.[38] Marx declared: "Freedom, in this sphere, can consist only in this, that socialized man, the associated producers, govern the human metabolism with nature in a rational way, bringing it under their collective control ... accomplishing it with the least

36 Saito, "Modern Agriculture," 25–30. As Engels wrote in 1876, capitalist production could not see beyond immediate profits. As an example, he described how plantations on Caribbean islands cleared forests without caring about the eventual consequences for the soil. Friedrich Engels, "The Dialectics of Nature," in Friedrich Engels and Karl Marx, *Karl Marx, Frederick Engels: Collected Works, Vol. 25* (New York: International Publishers, 1975), 463.

37 Karl Marx, *Capital*, vol. 1 (New York: International Publishers, 1967), 239.

38 See Foster, *Marx's Ecology*; Marx, *Capital: Vol. 1*, 860; Brett Clark and John Bellamy Foster, "Guano: The Global Metabolic Rift and the Fertilizer Trade," in *Ecology and Power*, eds. Alf Hornborg, Brett Clark, and Kenneth Hermele (London: Routledge, 2012), 68–82.

expenditure of energy and in conditions most worthy and appropriate for their human nature."[39]

Land tenure played an important role in Marx's thinking about the metabolic rift. Given their faith in scientific expertise and centralized planning, Marx and Engels insisted that small landowners could never preserve the planet's fragile soils on their own. Irrational, short-term fears prevented intelligent planning. Marx would have insisted that the solution to the problem of metabolic rift required collectivization and state planning to overcome the shortsightedness of individual farmers. Larger farms would simplify the process of planning and oversight. Moreover, small farmers were doomed to extinction as they struggled to compete with large landowners. They also played a reactionary role in society given their refusal to let go of their private property.[40]

Karl Kautsky, the nineteenth-century Marxist that most thoroughly explored the agricultural question, shared some of those assumptions about small landowners, but he challenged Marx nonetheless. Kautsky had doubts about whether the demise of small farming would occur exactly as Marx imagined. When writing about North America, Kautsky showed how the decline of soil fertility on prairie soils inevitably led to the decline of large landholdings. As long as virgin soils with fertile soil remained, investments in the soil could be ignored; after soils began to decline, however, small farms replaced large estates as they were better suited to the time- and labor-intensive practices of crop rotation and manuring that were required to revive depleted soils. In Europe, he noted that large estates did not have the same advantages as large factories. Thus, small holdings survived longer in the face of competition from larger estates.[41]

39 Marx, *Capital*, vol. 3 (London: Penguin, 1981), 754, 959.

40 See Marx, *Capital*, Vol. 1, Chapter 27; Marx and Engels, *The Communist Manifesto*; Tony Cliff, "Marxism and the Collectivization of Agriculture," *International Socialism* No. 19 (Winter 1964–5), Chapter 2, accessed June 1, 2018, <https://www.marxists.org/archive/cliff/works/1964/xx/index.htm>.

41 See Karl Kautsky, *Die Agrarfrage: Eine Übersicht über die Tendenzen des modernen Landwirthschaft und die Agrarpolitik der Sozialdemokratie* (Stuttgart: Dietz, 1899), 148–150, 298.

Even if he cast doubt on the short-term demise of small farms, Kautsky believed that small farms were not a long-term solution to the problem of the metabolic rift. In a compelling and insightful chapter, Kautsky illustrated how small families increasingly deprived their bodies and the soil to compete on the market against larger farms. They exhausted the soil to maximize revenues, or they sacrificed calories consumed by family members to increase revenues. They also mined the bodies of livestock by relying on fewer animals to perform a greater number of tasks, thus causing milk production to decline and dairy products to have less nutritional value. To the extent that small farms better preserved soil fertility, Kautsky concluded, they only did this by exploiting child labor and mining calories from other sentient beings. Paternal authority coerced the labor of women and children for the labor-intensive work of manuring fields and rotating crops. Religious dogma and conservative praise for tradition worked to retain this labor force in the face of the alluring freedoms of the city. Even if Kautsky admired small farmers who recycled nutrients and tended to soil health, he still insisted on liberating those women and children from this coercive regime through collectivization and cooperation.[42]

To a certain degree, Vladimir Lenin shared Kautsky's ambivalence about the fate of small landholdings. Lenin especially appreciated Kautsky's insights on small farmers and their mining of soil, livestock, and bodies to remain competitive. Lenin, for example, wrote, "Small production can hold on only because of the desperately reduced consumption and laborious, arduous toil. The dispersion and waste of human labor, the worst forms of dependence of the producer, exhaustion of the strength of the peasant family, of peasant cattle and peasant land – this is what capitalism brings to the peasant everywhere."[43] Lenin, as Kautsky did, also insisted that collectivization was the only solution. While the Bolsheviks did redistribute land to small landholders, they did so only out of necessity and not due to ideological principle.

42 Kautsky, *Agrarfrage*, 92–129.

43 Vladimir Ilyich Lenin, *Selected Works*, Volume 12 (Moscow: Progress Publishers, 1970), 248.

Lenin had long fought against Social Revolutionaries agitating for aid to small landholders.[44]

Pniower's work in 1948 owed a debt to these earlier works. As far as similarities are concerned, Pniower did foresee the ultimate end to the small farm and believed in the power of collectivization to better conserve natural resources. The most important similarity of Pniower's work with his predecessors, however, was his similar understanding of the metabolic rift. Agricultural planning had to consider the local and global cycling of nutrients to avoid degredation and decline. Pniower and Kautsky also both imagined that even a decentralized and less collectivized agricultural economy required large amounts of collaboration and state planning to become a reality.

Kautsky and Pniower both also envisioned a long transitional stage where small landholdings would continue to exist and would only dissolve gradually and democratically. Typically, critics of capitalism's metabolic rift identified large capitalist landholdings on the Great Plains as the premier example of waste and plunder. Large-scale farming lacked an intimate relationship with the soil and thus wasted it. Kautsky repurposed the term. For him, a type of socialist "latifundia," or estates, offered the best hope to retain rural farm labor and prevent the migration of young rural residents to the cities. He wrote, for example, "If new socialist latifundia were built and large estates no longer were worked by needy wage slaves, stable cooperatives of free, happy people would emerge [among the laborers] on those latifundia, and young people would no longer flee their small parcels for the cities. Instead they would flee the parcels for the large collective farms in order to escape the barbarism of the small family farm."[45] This vision, in part, resembled Pniower's later notion of a mixed urban/rural landscape. Peasants would no longer flee the countryside to find themselves working in large factories in the massive urban centers. Production centers could be distributed more evenly across the land, and the "murderous overcrowding" of the population in metropolitan centers would come to an end. They both therefore had a vision of a decentralized German economy with small and

44 Cliff, "Marxism and the Collectivization of Agriculture," Chapter 6.

45 Kautsky, *Agrarfrage*, 300.

middle-sized communities where industry and agriculture would live side by side. Such an arrangement would end the metabolic rift and gradually lead to a truly socialist agriculture.

While both Kautsky and Pniower criticized an overly centralized economy, they differed slightly in tone in how they described small farms. Pniower certainly did not have such a dismal view of labor exploitation on the family farm. Neither of them precisely defined the organization of property relations and collectivization in their imagined futures, so limited conclusions can be made. At the very least, Pniower saw promise in small farms that better cared for the soil than large mechanized estates. Kautsky, however, believed the exploitation of family labor overshadowed any benefits of such stewardship.

Long after Pniower's 1948 musings on small landholdings, Tony Cliff offered further evidence that socialist agriculture did not have to follow the exact path laid out by Marx or Lenin. The British socialist reflected on the continued survival of the family farm in the 1960s and expressed his impatience with the dictatorial collectivization practiced in most of Eastern Europe. His 1964 critique of collectivization demonstrated how socialist revolution actually gave (or would give) new life to the family farm.[46] Cliff did believe that a successful socialist revolution would ultimately undermine the private farm, but he saw this as a very long process. Initially, socialist reforms would make the life of the family farmer easier as markets would be stabilized, technology more readily available, and other inefficiencies of capitalist markets would end. The farmer would be able to manage his land without as much desperation and thus would not need to wring as much energy out of the soil and the bodies of his family and livestock. Cliff recognized the cultural appeal of land ownership, as it gave the rural poor or former proletariat workers a sense of autonomy and freedom in the opportunity to rule their own roost. As social welfare benefits and living standards generally increased, however, such autonomy would lose its allure. As Cliff concluded, "Capitalist pressure is not able to eliminate the small farm; socialist prosperity, by

46 Cliff, "Marxism and the Collectivization of Agriculture," Chapter 14.

attraction, will gradually – in the very long run – persuade the peasantry to give up their individual farms."[47]

East Germany's large collectivized farms were ideologically predetermined to a great degree, but perhaps not inevitable. Marx, Kautsky, and Lenin all identified inefficiencies in the small family farm. Either it could not compete with larger estates, wallowed in ignorance and tradition, or only survived through a patriarchal system that overworked and abused family labor to maintain solvency and fertility. Nonetheless, Kautsky's more nuanced analysis partially recognized the virtues of small farms for their ability to more carefully manage the soil and adapt to economic turbulence. In addition, followers of social democracy, such as Pniower, shared a widespread German belief that small farms benefited soil fertility and prevented the worst environmental degradation. Small farms also limited the political influence of Junker estates that Marxists tied to fascism and militarism. Read side-by-side with Kautsky, Pniower's sympathy for the small farm stands out. His skepticism about the long-term survival of the small farm appears less certain than Karn depicted it in her comparison of Pniower and Funcke.[48]

Biophysical Limits and the Postwar Context

In order to understand Pniower's friendlier attitude to small farms (when compared to previous socialist agricultural writings), the particular context of the late 1940s has to be considered. While an intellectual history of socialist agriculture can help reveal the similarities of Pniower's land

47 Cliff, "Marxism and the Collectivization of Agriculture," Chapter 14, second to last paragraph.

48 For more on German pride in small farms, see Scott Moranda, "'It makes no sense to work against nature': Cold War Modernization in West German Agriculture," in *Nature and the Iron Curtain*, eds. Astrid Mignon Kirchhof and John McNeill (Pittsburgh, PA: University of Pittsburgh Press, 2019).

reform proposals to older Marxist agricultural thinking, the distinctiveness of his text is best explained by the characteristics of that short period from 1945 (perhaps as early as 1943) to approximately 1950. Living in this period, Pniower, more than Kautsky, was acutely aware of scarcity and the limits of human innovation to solve food crises. Perhaps because of this, Pniower downplayed Kautsky's concerns about the use of family labor or the heavy reliance on livestock to manure fields. Rather than focus on the exploitative labor relations in that system, Pniower highlighted the benefits of family farms for soil and German self-sufficiency.

In the wake of Nazism, the western world's faith in technology's ability to sustain population growth faltered. Malthusian concerns about resource scarcity were not new, of course.[49] The war itself partly explains concerns about the degradation of resources, as warring parties plundered the planet to supply their war efforts, but the heightened sensitivity to Malthusian thinking had more to do with politics. The rise of fascism had made many defenders of democracy acutely aware of the need to better provide material comforts and caloric minimums to reduce the allure of extremist political movements. Anxious world leaders worried over how much more prosperity they could squeeze out of the land. Conferences began during the latter years of the war brought together food experts to address food distribution.[50]

Socialists retained their fundamental faith in centralized planning, science, and technology to raise the living standards of the working class, but their anti-fascism also made them aware of Germany's overconsumption of resources. According to them, fascism was a last gasp effort to defend

49 See Thomas Robertson, *The Malthusian Moment: Global Population Growth and the Birth of American Environmentalism* (New Brunswick, NJ: Rutgers University Press, 2012); Nick Cullather, *The Hungry World: America's Cold War Battle against Poverty in Asia* (Cambridge, MA: Harvard University Press, 2013); Amy L. Sayward, *The Birth of Development: How the World Bank, Food and Agriculture Organization, and World Health Organization Changed the World, 1945–1965* (Kent, OH: Kent State University Press, 2006).

50 *United Nations Conference on Food and Agriculture: Hot Springs, Virginia, May 18 – June 3, 1943: final act and section reports* (Washington, DC: Government Printing Office, 1943).

capitalist production, despite its inefficiency and waste. Marx's concern about a metabolic rift also continued to influence socialist agronomy. Concerns for metabolism at least raised questions about the sustainability of industrial systems and modern practices. In this context, socialists such as Pniower might have been more likely to reconsider the small farm.

The unique material conditions of postwar Germany also must have influenced Pniower's 1948 plan and similar ones in the West. With a complete collapse of the German economy upon military defeat, Pniower expected Germany to never regain its leverage over global markets that it would need to import feed for livestock or to fully provide calories to large urban centers. He predicted at the time that many workers might never return to industry as Germany's stature in the world declined, so small gardens and farms made sense as an alternative source of employment. In addition, Pniower imagined a postcolonial future where Germany and Europe would no longer be able to rely on the exploitation of tropical vegetable oils and proteins for their own benefit. Pniower thus recommended a reduction of the livestock herd and a change in German diets. While cattle helped farmers return fertility to the soil and provided vital proteins, dairy producers in Germany had continuously overstocked their herds and had become reliant on oils and protein imported from European colonies or from North America to feed their cows.[51] Germany also relied on those same distant regions for fat to feed workers. Instead of importing livestock feed, he proposed that farmers plant oilseeds and fruit to lessen the country's reliance on imports of these expensive items. A transition from large-scale grain farming and concentrated dairy production to small-scale vegetable and mixed dairy farming, he imagined, would benefit soil fertility, as small farms and gardens had always better tended the soil with compost, crop rotation, and other soil improvement measures.[52]

This window of heightened Malthusian panic transcended the East–West divide. In West Germany, Henry Morgenthau, Jr. imagined a German future remarkably similar to Pniower's. He insisted that Germany had been living beyond its means by exploiting distant resources. In part, he

51 See Karl Brandt, *The Reconstruction of World Agriculture* (New York: Norton, 1945).

52 Pniower, *Bodenreform und Gartenbau*, 10, 31–33, 39, 56–57, 63, 71.

intended to shrink the German economy to help it better fit its biophysical limitations.[53] This was not just proposed in order to punish Germany. Instead, he believed (as did many authorities in the occupation force) that Germany would no longer have any leverage in global trade and would need to grow more food at home. He framed this transition as a return to a more natural balance between agriculture and industry that would adjust Germany's consumption to the natural carrying capacity of its landmass.[54]

It is important to note that Pniower's (and Morgenthau's) consideration of a postcolonial future differed fundamentally from the economic plans that came to dominate West German planning offices in the 1950s. Western economic planners began to imagine a future where Germany regained the benefits of trade with colonial or postcolonial spaces. Karl Brandt, a liberal exile from Nazism, emerged as a major figure in postwar West German agricultural planning. In his writings, he depicted the global trading system required to support meat and fat consumption and industrial production in Germany as "natural" and unalterable. To sustain beef and dairy industries, farmers would continue to import concentrated feed materials imported from colonies or former colonies. Similarly, Carl Ross of the Food, Agriculture, and Forestry Group of the US Military Government proposed using German "know-how and organizational ability" to help develop plantations for "certain foods such as fats, oils and sugar [that] can be produced in tropical and subtropical areas much more cheaply than under a temperate climate."[55] Since the Allies would resist "any German overseas development that even remotely resemble[d] a colony,"

53 Henry Morgenthau, *Germany is Our Problem* (New York: Harper, 1945), 46–48, 62–64, 68–69.

54 The scholarship on the Morgenthau Plan is underdeveloped and some of it is highly politicized. John Dietrich, *The Morgenthau Plan: Soviet Influence on American Postwar Policy* (New York: Algora Pub. 2002); Warren F. Kimball, *Swords or Ploughshares?: the Morgenthau Plan for Defeated Nazi Germany, 1943–1946* (Philadelphia, PA: Lippincott, 1976); Jeffry M. Diefendorf, Axel Frohn, and Hermann-Josef Rupieper, *American Policy and the Reconstruction of West Germany, 1945–1955* (Washington, DC: German Historical Institute, 1993).

55 Brandt, *The Reconstruction of World Agriculture*, 16, 30, 48, 253, 369; Carl Ross, OMGUS, "Notes on Overseas Food Production Scheme for Germany, September 6,

he imagined 1.5 million hectares of plantations in Portuguese Angola or somewhere in South America developed by German experts "strictly along business like lines, preferably with US capital participating."[56] By reconnecting West Germany to colonial and neocolonial sources of animal feed and vegetable oils, western economists and agronomists could deflect proposals to restructure German agriculture.

While Pniower's interest in small farms and soil fertility reflected short term concerns of the immediate postwar period, he soon embraced mechanization and rapid collectivization in the 1950s. Heterodox ideas about mixed agricultural landscapes and small farms that survived within scientific institutes, landscape planning offices, and the Ministry of Agriculture became increasingly taboo. Authorities quickly tarred such plans as vestiges of an out-of-date and theoretically flawed Malthusianism. The strain of Marxist agricultural planning that stressed rapid industrialization and economies of scale dominated more than ever before. As authorities touted repeatedly, collectivization, fertilizers, mechanization, cybernetics and other scientific innovations guaranteed increased food production.

In part, Western strategists of containment provided fuel for orthodox Marxists to silence dissent. Western agronomists and politicians increasingly celebrated the small farm as a crucial weapon against the spread of communism. In their eyes, small farmers embodied the ideals of self-reliance, conservatism, soil care, and property ownership. American agricultural aid to Asia and Europe, therefore, sought to bolster the small farm and avoid a proletarianization of the countryside that might encourage radicalism. Eastern European leaders, in response, labeled any reform proposal with the slightest hint of "decentralization" as reactionary misinformation influenced by western propaganda.

Moreover, mechanization and fertilization soon boomed with the availability of cheap oil. An abundance of oil made a transition to synthetic nitrogen fertilizers and mechanized farms easier. Inexpensive oil,

---

1949"; General Correspondence (1933–71) Folder; Box 1; Henry G. Bennett Papers, Truman Library.

56 Brandt, *The Reconstruction of World Agriculture*, 1.

in other words, helped push aside earlier concerns about scarce resources and plunder. Technological solutions to food supply thus became predominant. In the West, agronomists began to think of improved methods, fertilizers, and specialization as a sort of "invisible colony" to supplement supplies of animal feed from postwar colonies.[57] In the East, oil and an admiration for heavy industry helped encourage the mechanization of collective farms. Soil mining no longer concerned agronomists given the growing abundance of nitrogen and other minerals through synthetic production.[58] In fact, farms in both the East and the West increasingly struggled with how to cope with an excess of nitrogen (rather than a shortage). Liquid manure and fertilizer runoff made worse by excess fertilizer application began to deteriorate water sources wherever industrial agriculture took root.[59]

Finally, the battle for political legitimacy pushed technocrats toward a crude vocabulary of tonnage and food quantity when it came to matters of agriculture. The East German government promised that socialism would provide better and more equal living standards for its citizens. Especially in the wake of the 1953 uprising, the regime pushed to avoid shortages of basic foodstuffs. The availability of abundant and inexpensive milk, meat, and potatoes became crucial to political stability. Letters of complaint and grumbling in media reports always increased if grocery stores, campgrounds, or restaurants failed to provide basic supplies. Collectivization and mechanization also helped eliminate sources of political resistance in the countryside.[60] The faith in science received a further boost with economic reforms in the 1960s when cybernetics and engineering solutions

57 Alfred Koenekamp, "Grünland – die unsichtbare Kolonie," *Deutsche Bauernzeitung* 2, no. 20 (1949), 5.

58 See Uekötter, "Why Panaceas Work"; Frank Uekötter, "Know Your Soil: Transitions in Farmers' and Scientists' Knowledge in Germany," in *Soils and Societies: Perspectives from Environmental History*, eds. John Robert McNeill and Verena Winiwarter (Isle of Harris: White Horse Press, 2006).

59 Frank Uekoetter, "Das flüssige Gold der Landwirtschaft," *Kuriosa der Wirtschafts-, Unternehmens- und Technikgeschichte: Miniaturen einer 'fröhlichen Wissenschaft'* (Essen: Klartext, 2008), 77.

60 See Last, *'Socialist Spring.'*

promised to help East Germany produce more chickens and sausages than ever before.[61]

Pniower's call for a garden landscape of small plots, in other words, very quickly became outdated. Since political and economic realities had changed at a breakneck pace after the war, Pniower's text seems today completely misplaced and outside of mainstream socialist thought. In reality, his work fit within certain traditions of socialist agronomy, and it embodied postwar beliefs about a new normal of scarcity for Germany. Reading Pniower's 1948 work side by side with earlier texts, it also becomes clear that his call for land reform shared with his socialist predecessors a condemnation of American plunder and a particularly Marxist call to heal a metabolic rift caused by urbanization and capitalism. Pniower joined Kautsky in considering the possibilities of the small farm, even if it seems that Pniower in 1948 had more faith in the sustainability of small farms than his predecessor. In the particular material and political context of the late 1940s, earlier revisions of Marx's theories of a disappearing peasant class may have gained more traction. Changes to socialist and West German agricultural policy after 1948 toward large-scale agriculture, it would seem, had as much to do with changing material and political realities after the war as it did with conflicting interpretations of Marxist doctrine.

At the end of the Cold War, the fall of the Berlin Wall symbolized to many the total failure of Marxist theories, and the story of the Soviet bloc's ecocide strongly linked Marxism's promethean instrumentalization of nature and crude materialism to environmental degradation and a collapse in public health. Until recently, Marx's concerns for nature received little or no attention. It was assumed that his attention to labor and the means of production did not make room for environmental concerns. Certainly, aspects of Marxist ideology encouraged a blind faith in science and technology (within a non-capitalist system) to transform nature for the benefit of the working class. Care, however, must be taken. Marxist

61 For an introduction to the history of consumer demand and economic planning in East Germany, see Eli Rubin, *Synthetic Socialism: Plastics & Dictatorship in the German Democratic Republic* (Chapel Hill: University of North Carolina Press, 2008), but also Poutrus, *Goldbroilers* (2002).

ideology was not monolithic, and certain material and political conditions encouraged some strains of the ideology while discouraging others. Many choices made by leaders of socialist dictatorships had much more to do with political survival and Cold War competition.

Today, many environmentalists celebrate the small family farm as an ecologically sustainable alternative to the factory farm. They might even turn to Marx's musings on metabolism to challenge corporate farming. Even as we have become more aware of Marx's writings on social metabolism, however, it is difficult to ignore Marx's disinterest in the small farm, which he believed had few redeeming qualities. Even Marxist agricultural reformers who revised Marx's thoughts on small farms paid little attention to their ecological virtues. Environmental collapse, in their eyes, was less likely when centralized planning helped science and reason triumph over self-interest and greed. Pniower's work is especially interesting in this light. While non-Marxist influences might have shaped Pniower's praise for the sustainability of small farms, the unique circumstances of defeated Germany must have played a role. For a brief moment, the western world (socialist and non-socialist) accepted that unending growth might not be possible, at least for the defeated countries. As the world today questions the ability of technology and science to prevent planetary disaster, perhaps socialist thinkers will again revisit the small farm and its fate.

TOBIAS HUFF

Environmental Policy in the GDR: Principles, Restrictions, Failure, and Legacy

There is, seemingly, little room for debate when comparing the environmental track records of the two Germanys. In the West there was a society and a government that took good care of its natural environment and cleaned up the pollution that had existed in its soil and air. The West German populace had a profound appreciation for the importance of the environment, and there existed in West German society a consensus in favor of making industry pay for cleaning up pollution, even if it meant higher consumer prices. And as a result, environmental laws in West Germany were highly effective in forcing real reforms on industry, leading to their significant investment in new technologies and facilities.

In the East, there was a party-controlled state and society whose economy was based largely on technology and machinery which was outdated, and so produced not only inferior goods but also pollution which seeped into the water, soil, and air. The pictures of biologically dead rivers, dying forests, landscapes with abandoned open pits or dirty and degenerated industrial complexes became emblematic for the environmental situation in the GDR. Especially in the 1980s, western media helped shape a narrative depicting a clear dichotomy between the "ecological West" and the ignorant East. The *Frankfurter Allgemeine Zeitung* – one of the leading West German newspapers – described the East German natural landscape in the direst terms, using terms like "a scene of destruction" and "wasted moonscapes."[1] Directly after the fall of the Berlin Wall,

1 Eckhart Kauntz, "Der Geruch verbrannter Braunkohle dringt überall hin," *Frankfurter Allgemeine Zeitung*, November 25, 1989; "Bitteres nicht nur aus Bitterfeld," *Frankfurter Allgemeine Zeitung*, November 8, 1988.

the newspaper described a scene in the small Saxon village of Schkopau: "In Schkopau there is no daylight. The workers, hurrying to the bus after the shift, seems to be like shadow beings on a planet far far away due to the smog dome. Lime powder lies on crippled trees, breathing is hard because of the stench. The opaque windows of the monster-like factory are burst, the brickwork is crumbling, smoke is seeping out. In an open cesspit, black sewage is slipping to the River Saale."[2] In early 1990 the magazine *Der Spiegel* ran the headline "Devil's Workshop" on its front cover, over an image showing five smokestacks in an Erlenmeyer flask. The corresponding article was entitled "Land of the Thousand Volcanoes."[3] From the western point of view the environment in the GDR was nearly destroyed, and its bad condition was interpreted as one reason for the collapse of the regime.[4] For so many in the West, the urgent need for environmental policy was motivated by apocalyptic visions of complete ecological collapse, as evidenced by the "forest death" (*Waldsterben*) debates and the reaction to the Chernobyl disasters. In East Germany, westerners saw that apocalyptic vision come to reality.[5]

After 1989, the question of how the environment in the former GDR came to be in such bad shape was not only the subject of media discussions; it was also the domain of scientific and scholarly discussion. One answer that surfaced in these early discussions was that the SED-Regime suppressed environmental thought and action out of fear that environmentalism could endanger the rule of the SED. The SED had always held

2 Axel Wermelskirchen, "Es wird Jahre dauern, bis der Dunstschleier über Halle zerreißt," *Frankfurter Allgemeine Zeitung*, November 25, 1989.

3 "Land der tausend Vulkane," *Der Spiegel*, January 8, 1990.

4 The more critical scientific evaluation of the environmental situation of the GDR had already begun in the mid-1980s and was not as one sided as the media. See Redaktion Deutschland-Archiv, ed., *Umweltprobleme und Umweltbewußtsein in der DDR* (Cologne: Verlag Wissenschaft und Politik, 1985).

5 See Kai F. Hünemörder, "Kassandra im modernen Gewand: Die umweltapokalyptischen Mahnrufe der frühen 1970er Jahre," and Patrick Kupper, "'Weltuntergangs-Vision aus dem Computer': Zur Geschichte der Studie 'Die Grenzen des Wachstums' von 1972," in *Wird Kassandra heiser? Die Geschichte falscher Ökoalarme*, eds. Frank Uekötter and Jens Hohensee (Stuttgart: Steiner, 2004).

that socialism believed in harmony with the environment, and that it was actually western capitalism that did not respect the needs of nature.[6] Thus, to talk openly about pollution would have been to call that belief into question, and so it needed to be repressed by the regime.

Of course, this western narrative is predicated on the assumption that there was, actually, an environmental awareness in East German society in the first place, one that could not flourish because of the power of the state to repress it. This interpretation of the fall of the GDR assumed that, free from socialist constraints, the GDR's people would naturally develop the same consensus regarding the need for drastic environmental policies as had happened in West Germany. It was a way of positioning the West German experience as the norm.

At first glance, this assumption seems reasonable. The 1989 revolution was incubated by grassroots environmental protest groups. These groups moved from protesting environmental pollution by sparring with the SED and state organs in the early 1980s to protesting basic human rights violations, such as the lack of freedom to travel or to speak openly about environmental and other issues. These ecological groups managed, in fact, to gain some room for maneuver due to two reasons. First, they were protected by the Protestant Church. And second, the SED could not persecute the groups to the full extent it wanted to because of how central environmental issues had become in West German mainstream culture and polity, and as a result, the western media already was watching the issue of environmentalism in the GDR closely.

As the SED began to lose its grip on power in autumn 1989, the networks of grassroots groups, so-called *Basisgruppen*, transformed themselves into more stable organizations, beginning with the New Forum (*Neues Forum*), called into being on September 9, 1989. Democracy Now (*Demokratie Jetzt*) was created on September 12, the Green League (*Grüne Liga*) on November 18 and the Green Party of the GDR (*Grüne Partei*

6 On this point see Joan DeBardeleben, *The Environment and Marxism-Leninism: The Soviet and East German Experience* (Boulder, CO: Westview Press, 1985) and Tobias Huff, *Natur und Industrie im Sozialismus: Eine Umweltgeschichte der DDR* (Göttingen: Vandenhoeck & Ruprecht, 2015), 187–191 and 250–252.

der DDR) on November 24.[7] As they formed the roundtable with some other groups, old parties and mass movement organizations, it was clear that color of the resistance to the SED was shaded green.

With meetings from December 7, 1989, onwards the roundtable influenced the policy of the new government, which had been formed after the resignation of Erich Honecker and was led by Hans Modrow, a reformist member of the SED. On February 5, 1990, these opposition groups entered into the cabinet, each with a minister without portfolio (Klaus Schlüte for the Green League and Matthias Platzeck for the Green Party). Environmental affairs had strong support in the cabinet, further rising when former dissident naturalist Michal Succow became deputy Minister for the Environment, Nature Protection and Water Management on January 14, 1990. One of the central efforts of the Ministry under Succow's leadership was to convert military grounds, especially restricted areas near the border, into nature parks. At the very end of its existence, the GDR suddenly strove to implement a flurry of environmental policy: on March 16, 1990 – just two days before the first free election in the GDR – the outgoing Modrow government sought to convert more than 10 percent of the territory of the GDR into national parkland, conservation areas and biosphere reserves.[8] And, the newly elected cabinet under Prime Minister Lothar de Maizière continued the work. Eventually, a legal ordinance was passed protecting 4.3 percent of the territory of the GDR. This was literally one of the very last things the GDR ever accomplished – the law entered into force on October 1, 1990 – two days before German reunification.[9]

7 Karsten Timmer, *Vom Aufbruch zum Umbruch: Die Bürgerbewegung in der DDR 1989* (Göttingen: Vandenhoeck & Ruprecht, 2000), 132–140 and 312.

8 Bundesarchiv (BArch) DC 20/I/3/2939, Ministerrat der Deutschen Demokratischen Republik, "Beschluß zur Information über den Stand und die vorgesehene Entwicklung von Biosphärenreservaten, Nationalparks und Naturschutzparks in der DDR (einschließlich Materialien)," March 15, 1990.

9 BArch DC 20/I/3/3063, Ministerrat der Deutschen Demokratischen Republik, "Beschluß zu den Verordnungen über die Festsetzung von Nationalparks sowie von Naturschutzgebieten und Landschaftsschutzgebieten von zentraler Bedeutung als Biosphärenreservate und Naturparks," September 12, 1990.

The great success for the eastern environmental movement continued with the first federal election of the reunified Germany on December 2, 1990 – in fact, the eastern Greens seemed more successful than the western Greens, because it was the East German electoral alliance *Bündnis 90* [Alliance '90], containing numerous environmental activists, which met the 5 percent threshold and made it into the Bundestag, while the western Green Party (*Die Grünen*) missed the electoral threshold and failed to gain any seats for the first time in years. This was due, in part, to the fact that in the FRG, security and economic issues took center stage in the months after the *Wende* of 1990, with the result being that during 1990–4 the *eastern* Greens were more successful than *western* Greens, who had, up until 1990, experienced a meteoric rise as the first new, viable party in the FRG since its founding. The face of the Greens in the Bundestag was no longer West German activists and politicians like Petra Kelly, but rather former dissidents like Klaus-Dieter Feige and Vera Wollenberger – one reason that to this day, the "Green Party" is officially titled "Bündnis90/Die Grünen."[10]

In the 1994 federal elections, however, this state of affairs came to an end. The Green Party failed to meet the 5 percent threshold in the "new federal states" of the former GDR, while it surged back to its previous totals in its original stomping grounds in West Germany, and to this day, for the most part, this situation has held – the Greens, despite the "Bündnis 90" in their name, are a West German party. In fact, in the most recent federal election in 2017, the Green Party missed the threshold in all of former East German states but one, Brandenburg, where it got exactly 5 percent of the vote, while passing it, in many cases with double digit results, in the western states. The same trend has held in state elections in the former East, where the Greens have met the threshold less than half the time.

The reality was that environmental and human rights groups in the GDR were well organized because they had to be in the face of the repression they had faced in the GDR, and as a result, they were in a good position to seize the opportunity that the *Wende* afforded them to put their causes at the top of the agenda. But as the *Wende* ran its course over several years, economic

10 Ludger Volmer, *Die Grünen: Von der Protestbewegung zur etablierten Partei* (Munich: Bertelsmann, 2009), 303–338.

issues became determinative for most former East Germans, and the issues that had motivated the citizens' groups who had sparked the 1989 revolution, such as human rights and the environment, fell into obscurity in the new federal states.[11] This was understandable – unemployment rose dramatically in the former East Germany, motivating voters in the East to turn to parties like the CDU, which promised economic reforms including currency unification, and away from the Greens, who did not campaign purely on issues of economic growth. Some authors have argued that the Green Party in the East was at a structural disadvantage as it did not have a preexisting party apparatus predating 1989, as did, for example, the CDU, though on further inspection, this argument does not hold up. First, the liberal FDP was able to use the preexisting party apparatus of the LDPD, and yet it too experienced the same disastrous results in 1994 (after an initial blush of success in 1990, like the Greens). Secondly, the eastern Greens may not have had a preexisting *East* German party apparatus, but they had strong support from the *western* Greens, who gave them computers, data, money and most importantly the know-how to run modern electoral campaigns.

No, the failure of the eastern Greens from 1994 to the present cannot be blamed on structural issues. Rather, it is the case that the prolonged weakness of the Green Party in the eastern part of Germany has to do with the legacy of the political, economic and social situation of the former GDR in regards to environmental issues, both before 1989 and afterwards. Based on the following examples I want to illustrate how this weakness resulted from a web of factors, including security questions, economic constraints, political calculations and social disinterest. The truth is that in the early years of the GDR, environmentalism was openly discussed, and the GDR was far "greener" than West Germany. Something went wrong during the 40 years that followed, and we still know relatively little about what that something actually was.

We are only beginning to really understand the environmental history of the GDR in of itself. Much of the German environmental history after 1945 has essentially been the environmental history of the FRG; the

11 Andreas Rödder, *Geschichte der deutschen Wiedervereinigung* (Munich: Beck, 2011), 42–48 and 58–63.

GDR is given cursory treatment in most general post 1945 overviews of German environmental history.[12] This could be a consequence of a lack of sources, as Jan Hünemörder has written.[13] Joachim Radkau has remarked that there is very little information about the Ministry of Environmental Protection and Water Management, a central protagonist in any environmental history of the GDR.[14] Of course, the lack of scholarly attention to GDR environmental history might also be a consequence of the pitiable state of the environment at the end of the GDR. Why bother writing a history of a failed environment in a failed state?

I argue it that it is illuminating to examine the environmental history of the GDR for multiple reasons. First, the dichotomy between the East and the West is not as black and white as it is held to be in popular perception, a perception that is in part a result of media coverage during the collapse of the GDR. Secondly, understanding what the history of the environment in the GDR actually helps us understand why environmental issues remained subordinated there to other concerns, and therefore helps us to understand a range of contemporary behaviors in Germany. In what follows, I explore the history of the environment in the GDR through the lens of both air pollution and the related issue of forest damage.

There are three reasons to focus on the issue of emissions-based air pollution as a window onto the environmental history of the GDR.

12 See David Blackbourn, *The Conquest of Nature: Water Landscape and the Making of Modern Germany*, 2nd edn. (London: Cape, 2006); Franz-Josef Brüggemeier, "Natur, Gesundheit, Eigentum: Zur Entwicklung des Umweltbewußtseins in Deutschland im 19. und 20. Jahrhundert," in *Schübe des Umweltbewußtseins und der Umweltrechtsentwicklung*, ed. Michael Kloepfer (Bonn: Economica, 1995); Franz-Josef Brüggemeier, *Tschernobyl, 26. April 1986 – die ökologische Herausforderung*, 2nd edn. (Munich: dtv, 1999); Michael Kloepfer, *Zur Geschichte des deutschen Umweltrechts* (Berlin: Duncker & Humblot, 1994); Christof Mauch and David Blackbourn, eds., *Nature in German History* (New York: Berghahn, 2004); Joachim Radkau, *Die Ära der Ökologie: Eine Weltgeschichte* (Munich: Beck, 2011); Frank Uekötter, *Deutschland in Grün: Eine zwiespältige Erfolgsgeschichte* (Göttingen: Vandenhoeck & Ruprecht, 2015).

13 Kai F. Hünemörder, *Die Frühgeschichte der globalen Umweltkrise und die Formierung der deutschen Umweltpolitik (1950–1973)* (Stuttgart: Steiner, 2004), 21.

14 Radkau, *Die Ära der Ökologie*, 139.

First, the field of emissions research and the study of forest damage in Germany largely originated at an internationally recognized research institute in the Saxon city of Tharandt. Secondly, emissions, especially of sulfur dioxide (largely responsible for acid rain), tended to become an international issue because of the nature of air pollution to travel across borders, and so by the end of the 1960s emissions became an issue frequently discussed in international diplomacy. The debate on acid rain, which was pushed forward by the Scandinavian countries, initiated a negotiation process which led to the 1979 *Convention on Long-Range Transboundary Air Pollution*. So, because of emissions in particular, the environmental policy of a state was no longer considered to be completely an internal affair; what policies states adopted within their borders affected neighboring states.

Thirdly, the *Waldsterbendebatte*, or "forest death debate," an intense and prolonged national discussion during the early 1980s about the potentially catastrophic effects of air pollution in the FRG, had a profound impact on both West and East Germany, and the ramifications of this "debate" still shape the ecological self-understanding of the Federal Republic. One of the ways in which this debate operated was that it brought out one major difference between the GDR and the FRG: fixing air pollution required technologies such as catalytic converters or flue-gas desulfurization plants that were expensive. The FRG could afford them, but they were largely beyond the means of the GDR to acquire. In this way, it came to appear that it was in fact free market economies which were better equipped to deal with industrial pollution than socialist economies. And the success that the West had, through legal and technological means, in preventing the feared "forest death" convinced many West Germans that they were living in a model ecological state.[15] I would like to add some grey to this largely black and white picture by taking a closer look at the some of the key historical actors within the GDR, by assessing how much agency they possessed within the constraints of the GDR state.

15 For more on German environmentalism see Uekötter, *Deutschland in Grün*, 9–10.

Reinhold Lingner and the "Landschaftsdiagnose"

"The whole organism of our landscape is sick," wrote the East German landscape architect Reinhold Lingner in September 1949 to the German Economic Commission (*Deutsche Wirtschaftskommission*). All economic activity in the country was, according to Lingner, being "carried out without regard for the consequences for the natural balance."[16] Lingner had been a convinced socialist since the 1930s and did draw inspiration from the success of Soviet planning.[17] Lingner's plans, however, differed markedly from the major projects in other state socialist environments. Like many planners and leaders in the Soviet world, he saw the natural world as fundamentally "malleable," yet he also renounced the destruction which was characteristic of some of the larger Stalinist landscape projects.[18] Lingner was aware that scientific and technological progress was a double-edged sword; it could serve the cause of human progress and simultaneously could lead to the human destruction of nature. In all his planning work, Lingner tried to stay true to the notions

16 "Der Abbau der Bodenschätze, die Gewinnung von Wasserkraft und Nutzung des Bodens wurde ohne Rücksicht auf die Folgen für den Naturhaushalt betrieben. [...] Der Gesamtorganismus unserer Landschaft ist krank." BArch DH 1/5725, Reinhold Lingner, "Präambel betreffend die von der Abteilung Landschaft im Rahmen eines Forschungsauftrages im Jahr 1950 vorgesehene Landschaftsbefundkartierung der Ostzone, September 6, 1949."

17 See Rüdiger Kirsten, *Die sozialistische Entwicklung der Landschaftsarchitektur in der Deutschen Demokratischen Republik: Ideen, Projekte und Personen – unter besonderer Berücksichtigung des Wirkens von Reinhold Lingner* (PhD dissertation, Hochschule für Architektur und Bauwesen, Weimar, 1989) and Kerstin Nowak, *Reinhold Lingner: Sein Leben und Werk im Kontext der frühen DDR-Geschichte* (PhD dissertation, Hochschule für Bildende Künste, Hamburg, 1995) for Lingner's vita.

18 Anselm Doering-Manteuffel, "Ordnung jenseits der politischen Systeme: Planung im 20. Jahrhundert," *Geschichte und Gesellschaft* 34 (2008): 399; also see Klaus Gestwa, "Technik als Kultur der Zukunft: Der Kult um die 'Stalinistischen Großbauten des Kommunismus'," *Geschichte und Gesellschaft* 30, no. 1 (2004): 45.

of "reconciling man with nature" found in the writings of Marx and Engels.[19]

Lingner was convinced that only a socialist planned economy could achieve this reconciliation, while the market economy based on short-term profits would lead to a "plunder economy" (*Raubwirtschaft*) and the overexploitation of natural resources. And Lingner's unique and nuanced viewpoint on the necessity of *both* socialist planning *and* care for the natural environment resonated with the GDR's leadership. In March 1950, the Ministry of Planning approved Lingner's "landscape diagnosis" project to the tune of 300,000 DM. In May 1950, 90 employees in five provincial groups set to work on the four tasks: 1) conservation of agricultural lands which did not have groves or thickets to serve as windbreaks and were thus exposed to erosion; 2) problems concerning water management; 3) disruptions to local culture as well as landscapes due to expanded mining operations; and 4) air pollution caused by industrial facilities, motor vehicles and dwellings emitting chemical agents.[20]

This ambitious plan of work ran right away into bureaucratic roadblocks, in particular concerning safety. In July 1950, the police in Thuringia forbade the mapping work. In August 1950, the work in the other areas was stopped. Behind at least some of the pushback was Hans Warnke,[21] State Secretary at the Ministry of the Interior, who wrote to the Ministry of Planning demanding to know whether the work of the landscape diagnosis project was necessary, since "the people involved are using large amounts of resources, and there is no guarantee the that the material is really being used

19 Friedrich Engels, *Umrisse zu einer Kritik der Nationalökonomie: Marx-Engels-Werke 1* (Berlin, GDR: Dietz, 1976), 505.

20 BArch DH 1/5725, Reinhold Lingner, Antrag auf Erteilung eines Forschungs-Entwicklungsauftrages an das Ministerium für Planung, HA Wissenschaft und Technik (AntragsNr. 430232/3290); Anschlußauftrag 15/21810, December 20, 1950.

21 Hans (Johannes) Warnke, born in 1896, former member of SPD and USPD was since 1920 a leading member of the KPD in Mecklenburg. During the Weimar Republic he was incarcerated on charges of high treason. During the Nazi regime he was twice imprisoned in Sachsenhausen. After the war he held leading positions both in the SED and the state administration. See Helmut Müller-Enbergs, ed., *Wer war wer in der DDR? Ein Lexikon ostdeutscher Biographien* (Berlin: Links, 2006), 1059–60.

solely for the purposes of building up our German Democratic Republic."[22] Here already we can see the conflict that Lingner wanted to transcend – pitting "building socialism" on the one hand against care for the environment on the other.

Investigations by the Interior Ministry revealed that some on Lingner's team had a National Socialist background, which further reinforced the suspicion among those in the Ministry that Lingner's attempt to "care for the environment" represented a project that was not aligned, or was even at cross purposes to, the building of socialism. On this point, the Central Commission on State Control (*Zentrale Kommission für Staatliche Kontrolle*) wrote that they were concerned about the mapping work Lingner was doing because the maps were "topographic maps" which could be militarily strategic, presumably to a foreign enemy, especially because they showed "air and wind conditions, especially in the areas of our industry."[23]

Still, eventually Lingner's work resumed, and in his final report, which was 1,119 pages long and included 951 maps, Lingner once again praised the socialist planned economy for allowing such a comprehensive project to come to fruition, writing "the complex investigation of the disturbance of the natural balance of the landscape as well as the planning of the landscape's rehabilitation were only made possible by the political conditions within the German Democratic Republic.[24]

These efforts resulted in Lingner's "Plan for the Implementation of the Transformation of Nature in Germany," which was discussed by the Secretariat of the Central Committee of the SED on June 1, 1953.[25] The "Plan" was to a degree based on the much more famous "Stalin Plan for the Transformation of Nature" in the USSR.[26] However, it is incorrect to

22 BArch DC 1/433, Hans Warnke, Forschungsauftrag "Landschaftsdiagnose der DDR," July 13, 1950.

23 BArch DC 1/433, Linz, Forschungsauftrag Nr. 15/21810 "Landschaftsdiagnose der fünf Länder der DDR als Unterlage der Landesplanung," March 29, 1951.

24 BArch DH 2/ 21697, Reinhold Lingner, Forschungsauftrag Nr. 240401-F2-03 "Landschaftsdiagnose der DDR," January 25, 1953.

25 BArch SAPMO DY 30/IV 2/7/94, Abteilung Landwirtschaft des ZK der SED, Plan für die Durchführung der Umgestaltung der Natur in Deutschland, April 24, 1953.

26 Reinhold Linger, *Landschaftsgestaltung* (Berlin: Aufbau-Verlag, 1952), 64.

simply argue that Lingner's plan therefore represented an East German version of Soviet gigantomania, as David Blackbourn has written.[27] Lingner's approach was far more cautious in his proposed interventions in the natural ecosystem, outlining how any projects undertaken should be based first on a careful observation of the ecological situation.[28] Lingner's Plan called for spending 15 million DM over the course of the next two Five-Year-Plans, and most of the funding was to go towards research, the establishment of sample landscapes, new research facilities and the establishment of an administration. The plan justified the need for the measures, and the large expenditures they entailed on the fact that both "Hitler fascism" as well as the reconstruction of the economy in the immediate postwar years had neglected the natural environment, and this needed to be remedied.[29] In a way, Lingner's vision was actually for a socialism based on small-scale agriculture. It was a vision of a socialist society and economy which would be sustainable. By knowing and understanding the details of the metabolic interaction, the required state of harmony between human beings and nature could be achieved, in his view.[30]

But the hardliners did not share his assessment; probably they did not even understand it. So, the concrete result of the deliberations in the Secretariat was therefore poor. The *Central Institute for the Transformation of the Nature* was not founded, and also the investigation of the landscape damage in the Senftenberg brown coal mining district did not begin. The Secretariat merely ordered two regulations to be drafted: one which would regulate the remediation of soil in areas with brown coal mining and a second protecting hedgerows in the agricultural landscape.[31]

27 Blackbourn, *Conquest*, 407–408.

28 Andreas Dix, "Nach dem Ende der 'Tausend Jahre': Landschaftsplanung in der Sowjetischen Besatzungszone und frühen DDR" in *Naturschutz und Nationalsozialismus*, eds. Joachim Radkau and Frank Uekötter (Frankfurt/Main: Campus, 2003), 352.

29 BArch SAPMO DY 30/IV 2/7/94, Plan für die Durchführung, 199.

30 Lingner, *Landschaftsgestaltung*, 7. For the postulate of harmony see Friedrich Engels, *Herrn Eugen Dühring's Umwälzung der Wissenschaft (Anti-Dühring): Marx-Engels-Werke 20* (Berlin: Dietz, 1962), 107.

31 BArch SAPMO DY 30/IV 2/7/94, Plan für die Durchführung, 4.

Lingner's Plan was to be derailed again, this time, not only by the Party hardliners but by the prelude to the uprising of June 16–17, 1953, which was directed against the SED's hardline policies. In particular, 1953 was sparked by the lack of consumer goods, including especially food, available for GDR citizens. East Germans were working hard to meet quotas set by the economic plan, but they were not seeing the rewards even in terms of basic necessities. In its wake, the GDR's leadership realized they needed to take drastic action to produce more consumer goods, including food, and they needed to do so quickly.[32]

This response to the crisis conflicted with Lingner's ideas presented above because in his plan food yields would increase sustainably – but slowly. In the light of 1953, that idea was anathema to the Central Committee and the Politburo, which sought to improve the supply situation immediately. The Party's leaders wanted large-scale agriculture and believed in a bold vision of imposing economies of scale onto natural lands, and in the wake of June 17, 1953, this became the Party's political line. In the aftermath of 1953, the SED leadership accused Lingner of "illusionism," saying that his plans were "romantic" and "oriented [...] to the visual appearance of untouched landscapes" and "did not fit into modern [...] social development."[33] Lingner resigned his position, countering that the Party leadership's accusations revealed a "disregard of natural laws" which would lead to a "significant weakening of the production power of nature." In his view, "the great possibilities offered by a socialist economy have not yet been utilized."[34] As his influence dwindled, he was replaced by another landscape architect named Georg Pniower, whose views of large-scale agriculture meshed much more closely with the SED's post 1953 exigencies.[35]

32 See Ilko-Sascha Kowalczuk, *17. Juni 1953* (Munich: Beck, 2013), 22–34 and Rolf Steiniger, *17. Juni 1953. Der Anfang vom langen Ende der DDR* (Munich: Olzog, 2003), 54–58.

33 Kurt Wiedemann, "Die Landschaft als Faktor der Planung," in Deutsche Bauakademie, *Landschaft und Planung*, 47.

34 Wiedemann, "Landschaft," 17.

35 Klaus-Dietrich Gandert, "Die Auswirkungen der Landschaftsdiagnose auf die anschließende fachliche Arbeit und Planungsvorhaben," in *Die Landschaftsdiagnose der DDR*, ed. Olaf Hiller (Berlin: TU Univ.-Bibliothek, 2002), 154.

Only after 1990 did Lingner receive public accolades[36] for his attempts to answer Marx's original call to transcend the "the antithesis between man and nature."[37]

Erich Zieger and the *Großraumdiagnose*

Erich Zieger was Professor of Forest Conservation at the forestry faculty in Tharand, part of the Technical University of Dresden. His research focused on a problem that had long bedeviled forestry experts, namely, the search for a means of protecting trees and woods against sulfur dioxide emissions. Since 1849 it had been widely known that sulfur dioxide emissions from factories caused damage to nearby forests. What was not known, and what Zieger's research sought to determine, was the question of at which concentration, and from which distance, sulfur dioxide caused its damage.[38]

Zieger was a supporter of Lingner's plan to use mapping to better understand damage to the landscape, meeting with him in 1955 in a working group dedicated to remediating pollution that was part of the State Planning

36 Nowak, *Reinhold Lingner*, 57; Helmut Gelbrich, "Landschaftsplanung in der DDR in den 50er Jahren," *Natur und Landschaft* 70 (1995): 543; Willi Oberkrome, "Sozialistische Heimat: Zum Natur- und Landschaftsschutz in der frühen DDR," in *Heimat: Konstanten und Wandel im 19./20. Jahrhundert*, ed. Katharina Weigand, Alpines Museum des Deutschen Alpenvereins – Schriftenreihe 2 (Munich: Bergverlag Rother, 1997), 235.

37 Karl Marx, *Ökonomisch-philosophische Manuskripte, Marx-Engels-Werke 40* (Berlin: Dietz, 1973), 536.

38 For the history of research on forest damage caused by smoke, see Svetlozar Rajanov, *Geschichte der Tharandter Immissionsforschung 1850–2002* (Dresden: Selbstverlag der Fachrichtung Forstwissenschaft der TU Dresden, 2002), and Otto Wienhaus and Hans-Günther Däßler, "140 Jahre Immissionsforschung am Institut für Pflanzenchemie und Holzchemie in Tharandt," *Staub – Reinhaltung der Luft* 51, no. 12 (1991).

Commission.[39] Zieger also became convinced that landscape planning, also called "spatial planning" or "large-area diagnostics" (*Großraumdiagnose*), could have great benefits for protecting forests, which in turn would lead to the reduction of economic losses.[40] In the centrally managed and planned economy, he regarded spatial planning as an instrument to reconcile the production interests of industry and forestry. In 1958 Zieger began working on a new research project that would make use of diagnostics for large land areas and that would be capable of linking measurements of air quality with damage to forests in one single report; this project received 219,000 DM in support from the Ministry of Forestry and Agriculture.[41]

Large-area diagnostics essentially yielded two results. First, statistical evidence showed that there was indeed a statistical correlation between airborne sulfur dioxide and the health of the forest, and second, the harmful effect of the pollution was much wider than had previously realized. The results of the diagnosis drew international attention; researchers from West Germany, Rome, Paris and Yale were interested in his results, in part because the research methods themselves were pioneering – demonstrating how innovative GDR science actually was at the time. Not only that, but one of Zieger's students who worked on the project warned of the possible catastrophic consequences of sulfur dioxide emissions in the Erzgebirge

39 BArch DE 1/4747, Staatliche Plankommission, "Entwurf einer Diskussionsgrundlage zum Komplex: Rauchschäden. September 19, 1955," 34–5.

40 BArch DF 4/60258, Institut für Forstnutzung, "Jahresbericht 1957 des Instituts für Forstnutzung der Fakultät füe Forstwirtschaft der Technischen Hochschule Dresden in Tharandt," January 20, 1958, 11.

41 BArch DF 4/52715, Institut für Forstschutz, "Jahresbericht 1961 der Professur für Forstwirtschaft und Forstschutz," 1962, 5; see also Herbert Lux, "Problematik und Methodik der Rauchschaden-Großraumdiagnose: Vortrag gehalten anläßlich der III. Internationalen Arbeitstagung forstlicher Rauchschadensachverständiger in Tharandt (24. bis 27. Mai 1961)," *Wissenschaftliche Zeitschrift der Technischen Universität Dresden* 11, no. 3 (1962) and Felix Lampadius, "Problematik und Methodik großräumiger Luftuntersuchungen: Vortrag gehalten anläßlich der III. Internationalen Arbeitstagung forstlicher Rauchschadensachverständiger in Tharandt (24. bis 27. Mai 1961)," *Wissenschaftliche Zeitschrift der Technischen Universität Dresden* 11, no. 3 (1962).

region, making the claim that within a short period of time air pollution would lead to large-scale deforestation.[42] This matter-of-fact presentation in the academic environment preceded the *Waldsterbenshysterie* [hysteria over dying forests] in the FRG by almost three decades.

Based on the results of his research, Zieger went on to advocate for restructuring industrial zones in ways that would cause a minimum of damage to surrounding forests. His proposals were rejected by the State Planning Commission because of the high costs they would have entailed due to interruption of production, relocation of plants and factories or the installation of filter technology. In addition, Zieger began in 1958 pushing for a law establishing a Central Commission on Smoke Damage (*Zentrale Rauchschadenkommission*) which would have placed limits on emissions.[43] Zieger was able to get a law to the drafting stage which would have placed stringent (for that era) limits on sulfur dioxide emission.[44] Yet, as before with Lingner, Zieger ran into resistance. Zieger found himself negotiating with the Planning Commission for two years, going through 10 versions of the bill, each time losing some of the restrictions on emissions he wanted until the language merely stated emissions should be held to "an acceptable level" as long as this limit did not overburden the economy.[45] Zieger

42 Horst Enderlein, "Welchen Einfluß haben die industriellen Emissionen auf das Kiefernsterben im Raum des Bezirkes Leipzig: Ergebnisse der durchgeführten Großraumdiagnose (Vortrag gehalten anläßlich der III. Internationalen Arbeitstagung forstlicher Rauchschadensachverständiger in Tharandt, 24. bis 27. Mai 1961)," *Wissenschaftliche Zeitschrift der Technischen Universität Dresden* 11, no. 3 (1962): 629–630.

43 BArch DK 1/3758, Ministerium für Land- und Forstwirtschaft, Verordnung Rauchschadenkommission, 126.

44 The suggested limits were 0.5 mg per cubic meter for short-term pollution and 0.15 mg per cubic meter for long-term pollution. BArch DK 1/3758, Ministerium für Land- und Fortswirtschaft, Verordnung über die Errichtung und den Betrieb von Anlagen zur Zurückhaltung von Staub-, Rauch- und Abgasschäden aus gewerblichen und industriellen Abgasen sowie die Diskussion darüber; Unterlagen zur Rauchschadenskommission, 1959, 52.

45 BArch DK 1/3758, Ministerium für Land- und Forstwirtschaft, Verordnung Rauchschadenkommission, 2.

blamed the powers of industry for stymieing him,[46] and when Zieger died suddenly in 1960, work on the law stopped.

While Lingner and Zieger worked on the subject of pollution, nature and industry in the 1950s, they differed somewhat in their approaches and philosophy. Lingner's vision was perhaps more holistic and long-term and attempted to combine lasting economic growth with the preservation of nature, while Zieger tried to solve a very specific, though fundamental, problem – that of sulfur dioxide emissions and the potentially catastrophic damage they were capable of causing to the forest. Both, however, believed that the centrally planned economy could be used to bring nature into harmony with industry. The example of these two pioneers clearly shows that the interests of industry were favored in conflicts between industry and nature conservation. At the highest political level, it was discussed at an early stage that the interests of nature conservation and landscape conservation should be taken into account in the plans, but given the supply situation immediately after the war and the beginning of the system competition, ecological concerns were systematically postponed.

The 1960s: Optimistic Mood and New Approaches

On May 29, 1968, the *Zentrales Vertragsgericht der DDR* [Central Contractual Court of the GDR], a court unique to the GDR that helped adjudicate between different industrial concerns in the planned economy, ruled that emissions were "accompanying phenomena of the scientific and technological revolution."[47] It ruled that, while it was not possible at that point to completely eliminate pollution, it was incumbent on industry to

46 BArch DK 107/8434, Deutsche Akademie für Landwirtschaftswissenschaften, Rauchschäden 1953–66, Brief von Ziege an Stubbe, 10.06.1959.

47 Staatsarchiv Chemnitz (StAC) 33209/623, Staatliches Vertragsgericht beim Ministerrat – Zentrales Vertragsgericht, Vertragsschiedverfahren StFB Dübener Heide gegen VEB Elektrochemisches Kombinat Bitterfeld, January 8, 1969, 9.

reduce the impact of such contaminants on the working and living conditions of the people. It acknowledged that the burdens of overcoming the devastation of the war, the consequences of a divided Germany, and the building up of a socialist economy had diverted attention away from dealing with pollution, but that by that point, in 1968, things were far enough along that reducing harmful emissions needed to be prioritized. Part of what allowed the Court to make this claim was its optimistic statement that the "economic potential" of the GDR had been growing in the last 20 years.[48]

It seems that the Central Contract Court put aside the circumstances that prevented Lingner and Zieger from implementing their plans: the undisputed primacy of economic concerns. Of course, this was not possible for the court to do alone; the court ruling should be seen in the context of the economic reforms which took place in the 1960s under Walter Ulbricht and known as the New Economic System of Planning and Leadership (*Neues ökonomisches System der Planung und Leitung* – NÖSPL).[49] In NÖSPL, the central planning apparatus was somewhat decentralized, giving individual companies more room for maneuver – but also bearing more responsibility, including responsibility for reducing pollution. Two examples highlight how this process worked in the GDR in the 1960s – the aforementioned landmark court case and an important research study.

On July 1, 1967, the State Forestry Company Dübener Heide sued the VEB (*Volkseigener Betrieb* [People's Own Factory], meaning a state-owned factory in the GDR) *Elektrochemisches Kombinat* Bitterfeld and 12 other companies in the District Court of Halle for damages totaling

48 Staatliches Vetragsgericht, 9.

49 Regarding the NÖSPL, see Gernot Gutmann and Hansjörg F. Buck, "Die Zentralplanwirtschaft der DDR: Funktionsweise, Funktionsschwächen und Konkursbilanz," in Kuhrt, *Die wirtschaftliche und ökologische Situation der DDR in den 80er Jahren*; André Steiner, "'Umfassender Aufbau des Sozialismus' oder 'Anleihe beim Kapitalismus?' Zur Darstellung des Konzepts der DDR-Wirtschaftsreform," in *Propaganda in Deutschland: Zur Geschichte der politischen Massenbeeinflussung im 20. Jahrhundert*, ed. Gerald Diesener (Darmstadt: Primus, 1996) and André Steiner, *Die DDR-Wirtschaftsreform der sechziger Jahre: Konflikt zwischen Effizienz- und Machtkalkül* (Berlin: Akademie Verlag, 1999).

7,303,010 marks due to emissions that had harmed the forest.[50] Shortly thereafter, the GDR's Central Contractual Court took up the case, citing its "economic importance for the future regulation of emission problems."[51] Bitterfeld denied any link between its emissions and the damage to the forest. However, the court came to a different conclusion, basing its judgment on the Land Use Ordinance (*Bodennutzungsverordnung*) from 1964 as well as a decision of the Supreme Court of the GDR from 1965, both of which had been created as parts of the economic reforms that Ulbricht had pushed through in the early to mid-1960s known as the "New Economic System" (*Neues Ökonomisches System der Planung und Leitung* – NÖSPL). The Land Use Ordinance had been amended in 1968 to give compensation to any agricultural interest that had its land use "restricted,"[52] and it was on this basis that the court found factories liable for damages to the landscape.[53] The Central Contractual Court ordered the Bitterfeld Combine and six of the 12 other companies named in the suit to pay 1,512,527 marks in compensation.[54] And yet, the court did not order the plant to take measures to reduce the sulfur emissions, with reasoning that the technology for such reductions either did not exist or was too expensive.

From an ecological point of view, the judgment was a double-edged sword. On the one hand, polluters were protected from having to stop their operations or retrofit filter technology. However, they were liable for the damages their pollution caused. Ideally, the cost of the litigation and settlements they would have to pay out would persuade companies to invest

50 StAC 33209/623, Vertragsschiedverfahren, 4. After 1967 the term "marks" is used for East German currency.

51 StAC 33209/623, Vertragsschiedverfahren, 4.

52 Ministerrat der Deutschen Demokratischen Republik, "Verordnung zum Schutz des land- und forstwirtschaftlichen Grund und Bodens und zur Sicherung der sozialistischen Bodennutzung – Bodennutzungsverordnung," *Gesetzblatt DDR Teil II (December 17, 1964)*, 233.

53 Der Vorsitzende des Landwirtschaftsrates der Deutschen Demokratischen Republik "Erste Durchführungsbestimmung zur Bodennutzungsverordnung – Ausgleich der Wirtschaftserschwernisse," *Gesetzblatt DDR Teil II (May 28, 1968)*, 295; StAC 33209/623, Vertragsschiedverfahren, 12.

54 StAC 33209/623, Vertragsschiedverfahren, 15.

proactively in retrofitting anti-pollution technology. But the reality was that acquiring and installing the technology necessary to reduce pollution remained more expensive than paying out claims for damages, or paying fines, and so most companies simply made a simple cost-benefit calculation to continue polluting. For example, decommissioning the boiler system at the enormous chemical factory complex VEB Leuna-Werke to retrofit for pollution reduction alone was estimated to cost 1 million marks.[55] Overall, at one point in the 1980s, the Ministry of the Environment calculated the investment needs for filter technology at 800 million marks – *a year*.[56] This is one reason that pollution continued in the GDR. The theoretical long-term advantages of environmental management in a planned economy, as Lingner once pointed out, could not be realized in such a scenario.

Towards a State Environmental Policy

In 1966 just before the Bitterfeld case, the Council of Ministers passed the "Decree on Measures to Prevent Air Pollution."[57] It defined 48 different airborne contaminants which needed to be regulated. Emissions were no longer just a civil court matter between the wrongdoer and the injured but also part of the state legislation. Since then, emitting companies had to pay compensations and fines in case of violation.

At the same time, 21 scientists were working on a fundamental study on environmental policy, entitled "Prognosis of Industrial Waste Products

55 Günter Streibel, "Erfahrungen mit planerischen Instrumenten der Umweltökonomie," in *Umweltpolitik in der DDR. Dokumente des Umbruchs*, ed. Arnim Bechmann (Berlin: Univ.-Bibliothek der TU, 1991), 39.

56 BArch DK 5/1355, Staatliche Umweltinspektion, Bericht zur Entwicklung der Emission von Luftverunreinigungen in der DDR im Jahre 1986. Berlin (Ost). 1986, 4–5.

57 BArch DC 20/I/4/1412, Ministerrat der Deutschen Demokratischen Republik, "Beschluß über Maßnahmen zur Reinhaltung der Luft – Auszug," October 1, 1966.

and the Planned Organization of Socialist Land Improvement in the GDR."[58] They came from a wide range of disciplines (lawyers, agricultural economists, natural scientists) and from universities, companies and research institutes. The group was led by Werner Titel, who was to become Minister of the Environment. The aim of the study was to develop an "economic experiment," which would demonstrate "solutions for air and water pollution." The 118-page study considered the problem of pollution as a complicated and interrelated set of causes and effects, stating that: "The losses in the economy due to damages, additional expenses and defaulted resource recovery reach the region of two billion marks annually in the GDR. To this must be added the harmful effect on the life-environment of humans, a substantial impairment of the hygienic conditions as well as the disfigurement of the nature, the impact of which is hard to fathom."[59]

One solution that the study suggested was the pricing of environmental resources. The "Permanent Working Group on Socialist Land Improvement in the Council of Ministers of the GDR" – convened in February 1969 – tackled this idea. The outcome of the deliberations was the "Act on the Planned Organization of the Socialist Land Improvement in the German Democratic Republic," passed on May 14, 1970.[60]

The "Landeskulturgesetz," as this law was known, was a landmark law; it was the second environmental protection law passed in Europe (Sweden had the first) and was very well received in the West, where one study published in 1973 praised the law as establishing "very progressive legal norms tailored to the conditions of a modern industrial society, with which a rich set of instruments for the protection of the environment had

58 BArch DC 20/I/3/715, Ministerrat der Deutschen Demokratischen Republik, Beschluß über die planmäßige Entwicklung einer sozialistischen Landeskultur in der Deutschen Demokratischen Republik, February 5, 1969.

59 BArch DC 20/I/3/715, Beschluß über die planmäßige Entwicklung, 45.

60 Der Vorsitzende des Staatrates der Deutschen Demokratischen Republik, "Gesetz über die planmäßige Gestaltung der sozialistischen Landeskultur in der Deutschen Demokratischen Republik (Landeskulturgesetz)," *Gesetzblatt DDR Teil I (May 14, 1970)*, 67.

been created."[61] Further environmental legislation followed, including the "Air Pollution Prevention Ordinance" of January 17, 1973, which fixed the emissions limits. Each company received an official certificate of compliance with emissions standards. If factories surpassed those standards (in either emitted dust or exhaust fumes), they were fined.[62] In the beginning, the Regional Sanitation Department (*Bezirks-Hygiene-Insepektionen*) supervised the limits; later the State Environmental Commission (*Staatliche Umweltkommission*) at the Ministry of the Environment, founded in 1985, took over this task.[63]

In 1976, the total volume of those fines was 7.5 million marks.[64] The money flowed into the rehabilitation of recreation facilities, spa holidays for families from contaminated areas or the expansion of medical care for citizens affected by pollution. The dust and exhaust fines were like an eco-tax, which was supposed to have influenced the investment behavior of the management. In a way, this was an experiment in pricing resources, and the fines themselves were quite innovative in this regard; they essentially increased the price of natural resources like water, soil and air, which previously could be used freely or almost freely in the production process. But the fines were, generally, too low to persuade the industry to install filter systems. The total combined amount of the fines per year never amounted to more than 47.3 million marks (in 1986).[65] In that same year, the cost of

61 Hans-Hermann Höhmann, Gertraud Seidenstecher, and Thomas Vajna, *Umweltschutz und ökonomisches System in Osteuropa: Drei Beispiele: Sowjetunion, DDR, Ungarn* (Stuttgart: Kohlhammer, 1973), 116–117.

62 Ministerrat der Deutschen Demokratischen Republik, "Fünfte Durchführungsverordnung zum Landeskulturgesetz (Reinhaltung der Luft), 5. dvo," *Gesetzblatt DDR Teil I (January 17, 1973)*, 157.

63 Anonymus, "DDR-Umweltinspektion – welche Aufgaben?" *Was und wie: Informationen, Argumente, Übersichten für den Agitator* 11 (1985): 16.

64 Ministerium für Gesundheitswesen, *Bericht der Abteilung und Hauptinspektion Kommunalhygiene über die Situation auf dem Gebiet der Lufthygiene, der Wasserhygiene, der Bodenhygiene und des kommunalen Lärmschutzes im Jahr 1976* (Berlin, 1977), 48.

65 BArch DK 4/1355, Staatliche Umweltinspektion, Bericht zur Entwicklung der Emission von Luftverunreinigungen in der DDR im Jahre 1986, 1986, 4–5.

a flue gas desulfurization plant at the Leuna power plant was estimated at 1.9 billion marks.[66]

The 1960s brought as well an expansion of normative regulations in environmental law as an expansion of economic incentive mechanisms. Environmental protection was supposed to be enforced through market mechanisms. Like much else with the NÖSPL, the reforms failed because they seemed to merge the worst of both systems. On the one hand, companies did not have to seriously contend with the threat of going bankrupt in a planned economic system, and so they did not have as much motivation to comply. On the other hand, the prices of environmental policy were politically determined – they depended on the willingness of the political leadership to protect the environment. This led to an inherent conflict of competing interests which would shape the environmental policy of the GDR until its end.

The 1970s: Environment Policy under Honecker

Erich Honecker's ascent to power in May 1971 coincided with the high point of GDR environmental policy. In 1972 the Ministry of Environmental Protection and Water Management (*Ministerium für Umweltschutz und Wasserwirtschaft* – MUW) was established. Between 1971 and 1973, a state initiative called the "Weeks of Socialist Landscape Care" (*Wochen der Sozialistischen Landeskultur*) began a mass information campaign concerning environmental issues. Nationwide there were events where authorities and companies could present their environmental efforts. Radio and television programs informed how people themselves could contribute to the protection of the environment. At the VIII SED Congress in May 1971, Honecker stressed the importance

66 BArch DK 5/2126 Werner Krolikowski, Bericht zur Entwicklung der Emissionen von Luftverunreinigungen in der DDR im Jahre 1986.

of environmental protection, and this was a high enough profile forum that it caught the attention of the West.[67] The environmental policy concepts, however, all originated earlier when Walter Ulbricht was in charge, and Honecker had no interest in further developing them. One example is the Ministry of Environmental Protection and Water Management, whose first task was education. It had only a small apparatus to inform people about environmental correlation or to implement and monitor legal regulations. The "Weeks of Socialist Landscape Care" were discontinued in 1974, and the corresponding annual report on the environmental situation became a classified document.[68]

Given that Honecker's environmental policies coincided with the Stockholm Environmental Conference in 1972, it seemed that the SED regime was using environmental policy to bolster its reputation abroad, but the reality was different.[69] In truth, work on the *Landeskulturgesetz* of the GDR began in 1967 with the aforementioned *Prognosis*. 1967 was a year before the initial preparatory work for the Stockholm conference began. It was already 1968, when the Economic and Social Council of the United Nations asked the UN General Assembly to convoke a conference to discuss the problems of the human environment.[70] This request could be seen as the starting signal for the environmental conference. Nonetheless, Honecker was no pioneer of environmentalism. He inherited a progressive environmental policy from his predecessor and used it as long as it seemed opportune, but he did not attempt to improve on the progressive environmental legacy he inherited.

67 Sozialistische Einheitspartei Deutschlands, *Protokoll der Verhandlungen des VIII. Parteitages der Sozialistischen Einheitspartei Deutschlands: 1. bis 3. Beratungstag*, vol. 1 (Berlin, GDR, 1971), 63.

68 Horst Paucke, *Chancen für Umweltpolitik und Umweltforschung: Zur Situation in der ehemaligen DDR*, Forum Wissenschaft/Studien 30 (Marburg: BdWi-Verlag, 1994), 41.

69 Radkau, *Die Ära der Ökologie*, 139.

70 UN Economic and Social Council Resolution 1346 (XLV), *Question of convening an international conference on the problems of human environment*, E/RES/1346(XLV) (July 30, 1968), available from <https://documents-dds-ny.un.org/doc/UNDOC/GEN/NR0/762/44/IMG/NR076244.pdf?OpenElement>.

The lack of support for environmental policy at the highest levels of the SED was reflected in the structure of the SED Politburo Central Committee. Often, the Central Committee would create its own department to mirror certain ministries or other important areas of policy and to concentrate the efforts of Party members. Frequently, the areas chosen by the Central Committee reflected their priorities. The Central Committee never created a subcommittee on Environmental Protection, and by the end of the GDR, the responsibility for environmental policy was still assigned to the Secretariat for the Economy, Department of the Basic Industry. As a result, the Minister of the Environment during the Honecker era, Hans Reichelt, struggled to find a connection with someone in the leadership of the SED. Thus, Reichelt found it very difficult to fulfill his goal of integrating economic concerns with the reduction of air pollution as part of the broader goal of environmental protection.[71] His Ministry of Environmental Protection was, at the level of the Politburo, dependent on the benevolence of Günter Mittag, who was the head of the Economic Bureau of the Central Committee. In the face of the economic and financial situation, Mittag paid less and less attention to ecological concerns. Short-term profits were more and more important.[72]

In fact, the bureaucratic structure of the Politburo reflected a general theme of prioritizing economic issues over environmental protection. This was made clear by the central platform of Honecker's regime, the "Unity of Economic and Social Policy," which was a series of policies that sought to increase the standard of living, or the quality of life, for East Germans as consumers. It did so through debt-financing. Going into debt to finance consumption instead of production had been considered a cardinal sin of

71 Ministerrat der Deutschen Demokratischen Republik, "Fünfte Durchführungsverordnung zum Landeskulturgesetz (Reinhaltung der Luft), 5. dvo," *Gesetzblatt DDR Teil I (January 17, 1973)*, 158.

72 Wolfram Bickerich, Dieter Kampe, Steffen Uhlmann, "'Es reißt mir das Herz kaputt': SPIEGEL-Gespräch mit dem ehemaligen DDR-Wirtschaftslenker Günter Mittag über seine Politik und seine Fehler," *Der Spiegel* 37, September 9, 1991, 88–104. See also "Zentraler Runder Tisch der DDR, Information zu den Ursachen der bisherigen Nicht-Umweltpolitik der DDR. Information Nr. 10/8," in *Umweltpolitik*, ed. Bechmann, 87.

central planning until that point.[73] Yet, Honecker was betting on a kind of supply-side state socialism, in which happier citizens would be more productive in the long run. At first, it seemed to work; although foreign debt skyrocketed within one year (by 43 percent),[74] socialism was perhaps never more popular than in the early years of the Honecker's government (1971–5).[75] This program, however, meant that, between expenditures for things like the massive Housing Program and debt servicing, there was little money left over for the protection of the environment.[76] The GDR was caught in a no-win situation because one of the reasons it was trying so hard to increase the standard of living was to compete with the FRG in terms of productivity and living standards, and yet it was not able to do so without sacrificing and cutting corners that would affect living standards, such as local environmental conditions.[77]

Honecker's bet came up empty in 1981. The hoped-for increase in labor productivity had ceased, the debts continued to grow, and on top of this, the USSR cut its oil supplies to the GDR by almost 2 million tons a year because it found out that the GDR was selling some of the oil to world markets in the West in order to obtain foreign exchange for servicing its debts and purchasing consumer goods.[78] Suddenly, to save resources and money, the GDR had to switch from petroleum as an energy source, as well as a source for synthetics,

73 Gareth Dale, *Popular Protest in East Germany, 1945–1989* (London: Routledge, 2005), 59.

74 André Steiner, *Von Plan zu Plan: Eine Wirtschaftsgeschichte der DDR* (Bonn: Bundeszentrale für Politische Bildung, 2007), 191.

75 Walter Friedrich, "Mentalitätswandlungen der Jugend in der DDR," *Aus Politik und Zeitgeschichte* 40, no. 16/17 (1990): 26.

76 For the importance of the housing program for the satisfaction of GDR citizens with the regime see Eli Rubin, *Amnesiopolis. Modernity, Space, and Memory in East Germany* (Oxford: Oxford University Press, 2016).

77 Beatrix Bouvier has vividly described how the unity of economic and social policy has increasingly narrowed the scope for investment. Beatrix Bouvier, *Die DDR – ein Sozialstaat? Sozialpolitik in der Ära Honecker* (Bonn: Dietz, 2002), 68–97.

78 Dierk Hoffmann, "Ölpreisschock und Utopieverlust: Getrennte Krisenwahrnehmung und -bewältigung," in *Das doppelte Deutschland: 40 Jahre Systemkonkurrenz*, eds. Udo Wengst and Hermann Wentker (Berlin: Links, 2008), 229–30.

to brown coal, or lignite, which burned much "dirtier" than petroleum.[79] Reverting back to lignite meant a stark climb in the amount of sulfur dioxide emitted in the GDR; it rose from 3.8 Mio. Tons of sulfur dioxide in 1973 up to 5.5 Mio tons in 1987. This was an increase of 45 percent (as well as a large increase in open-pit lignite mining, which caused environmental problems of its own).[80] 1981 was the year in which the "forest death" debate began in the FRG, and it was this debate which led to a massive reduction of sulfur dioxide emissions in the FRG. 1981 also was the year in which sulfur dioxide emissions began to rise as lignite usage began to climb due to oil shortages in the GDR became a starting point for unprecedented emissions in the GDR.[81] Between 1980 and 1984 alone, the production of lignite increased by 10 percent. With the bulk of the GDR's funds tied up in social policy and debt, Honecker had left the GDR with little room to maneuver in the absence of the support he needed from the Soviet Union, and thus it was environmental protection that had to pay the price.

Conclusion and Outlook

From the very beginning, the GDR was overwhelmed: reconstruction, preservation of power, reparations and the sincere attempt to improve the social situation of the population overstretched the limited possibilities.

79 The calorific value of lignite is significantly lower than that of hard coal or oil. Therefore, larger quantities must be burned. In addition, the sulfur content is much higher. But this too varies greatly. In the GDR, there were mining areas with 0.6 percent sulfur content and those with 1.6 percent. BArch DY 30/2836, Informationen zum Stand der Abgasentlastung, March 10, 1983, 1.

80 BArch DK 5/1961, Institut für Wasserwirtschaft, Die Entwicklung der Schwefeldioxidbelastung und ihre volkswirtschaftlichen Auswirkungen auf die DDR, January 31, 1977, 10; BArch DK 5/1306, Ministerium für Umweltschutz und Wasserwirtschaft, Bericht zur Entwicklung der Emission von Luftverunreinigungen in der DDR im Jahre 1987, 2.

81 On this point see Huff, *Natur und Industrie*, 191–200.

The example of Lingner, however, shows how the GDR leadership openly considered alternative approaches that thought out ecological issues right from the start. On June 17, 1953, however, the SED had to learn painfully that the supply of goods had the highest priority for the population. All other claims had to stand behind it, including environmental policy. The promising approaches of the 1960s, such as experimenting with innovative models to internalize external pollution, were abandoned in favor of an overarching transfer policy. The change of power in May 1971 can be seen as a decisive turning point in the GDR's environmental history. The expansion of social transfers in the 1970s served to stabilize and stimulate the economy. The concept of the "unity of economic and social policy" fixed the GDR into a path in which, faced with a choice between environmental protection and consumer spending, the choice was mostly the latter.

For about 35 years, the Party had relative success with this approach because the great majority of people in the GDR shared the focus on material goods, as opposed to the environment. However, when environmental issues became more and more important in the international context, the GDR had no room for maneuver to react. This is clearly evident in the negotiations on the Geneva "Convention on Long-Range Transboundary Air Pollution" from the late 1970s onwards. The GDR lacked the technology and means to reduce emissions and implement the agreements, leaving it with only one way out: to fake the data, with fatal effects on its credibility. The West German debate on supposedly dying forests in the 1980s especially revealed the shortcomings of GDR environmental policy and gave the independent environmental groups more room for maneuver.

ASTRID MIGNON KIRCHHOF

Counterworlds: The Pioneers of Nature Conservation and Life Reform in East Germany

Introduction

Erna and Kurt Kretschmann can be considered the pioneers of nature conservation in East Germany and held various honorary positions in official nature conservation structures. However, after 1960 they largely removed themselves from involvement in official structures and built a nature center called *Haus der Naturpflege* [House of Nature Care] on a leased parcel of land in Bad Freienwalde (in the federal state Brandenburg, in the northeast of the country). Here they were able to lead lives of material self-sufficiency; in addition, they organized a public nature and cultural center. For several decades they hosted guests from all over Europe with whom they exchanged ideas about nature conservation and built up networks. With a philosophy of life that built on the ideals of the *Lebensreform* [Life Reform][1] movement of the nineteenth century, the Kretschmanns created a counterworld[2]

1 Life Reform was a social movement in the late nineteenth century and early twentieth century that was a reaction to modern developments like industrialization and urbanization. It propagated a back-to-nature lifestyle, emphasizing, among other things, health food, dress reform, progressive education, nudism, sexual liberation, alternative medicine, rural settlement and abstention from alcohol, tobacco, drugs, and vaccines. Political ideas and ideologies within different groups reached from anarchism and socialism to conservativism and National Socialism, see Diethart Kerbs and Jürgen Reulecke, eds., *Handbuch der deutschen Reformbewegungen 1880–1933* (Wuppertal: Hammer, 1998).

2 Counterworlds are defined as spaces of action constituting a distinctive worldview which deviates from or contests hegemonic social or political structures. Of

that was in noticeable contrast to the lifestyle promoted by the Socialist Unity Party. This counterworldly sphere of activity, in which values and ideals could be negotiated and passed on to others, was based on a collective conviction that nature and environmental protection were vitally necessary and that a life in harmony with nature was both possible and essential. Erna and Kurt Kretschmann adhered to a Life Reform-inspired worldview centered on the individual and the individual's freedoms. They believed that individual-led reform was more effective than state-led reform because social change is driven primarily by insightful, committed, and responsible people. However, since according to Marxist thought reforms originate with the state and not with the individual, in East Germany the focus was placed on ways of living that were guided by the state. Official nature conservation and environmental policies also followed these Marxist maxims. Consequently, Erna and Kurt Kretschmann were faced with a difficult dilemma: they supported the socialist state in principle, but were at conflict with its approach to nature conservation. Furthermore, the East German system relied on the Party steering and controlling every possible aspect of state and society. Anyone who questioned this control and the supposed stability it achieved became a threat. How can we explain the non-conformist lifestyle of the Kretschmanns under such conditions?

---

fundamental importance for the genesis and development of counterworlds is, first, how the active players see themselves and how others see them. The inside and outside perspective may agree about the potential for disruption and innovation, but sometimes they diverge considerably: the novel character of certain ways of seeing and acting often only becomes clear in and through the reactions of others. Second, the development and perpetuation of alternative patterns of thinking and acting is dependent on the possibilities and limitations of exploring new social spaces or reallocating existing ones. These spaces are typically connected with actual physical locations, but are by no means always identical. Social spaces can also, for example, be opened up through discourses in the form of texts. I developed the definition used here in conjunction with sociologist Nina Leonhard; it draws on related sociological concepts such as social worlds (Strauss), subcultures (Williams), and counter-cultural spaces (Löw). See the introduction, "Gegenwelten," in *Geschichte und Gesellschaft* 41 (2015), 5–16.

How was it possible that in a dictatorial (*durchherrscht*)[3] society like East Germany a separate universe could be created, a world outside of the official structures that existed for decades in spite of close supervision by the Socialist Unity Party and the State Security?

Through their philosophy and practice, I argue, Kurt and Erna Kretschmann established a counterworld in opposition to the worldview of the Socialist Unity Party. This counterworld had a certain impact through a minority of people in East Germany who practiced Life Reform ideals, doctors who used naturopathic methods in their practices, and individuals who adopted a vegetarian lifestyle. The Kretschmanns became pioneers whose ideals and tenets were later taken up by parts of the East German environmental movement in the 1980s. While most scholarly literature focuses on the significance of Western influences on the emergence and subsequent path of this movement, in this article I want to draw attention to the significance of intellectual precursors to the later environmental groups within East Germany.[4] In the early 1980s the environmental movement had its onset in the German Democratic Republic (GDR), but the

3 Jürgen Kocka, "Eine durchherrschte Gesellschaft," in *Sozialgeschichte der DDR*, eds. Hartmut Kaelble et al. (Stuttgart: Klett Cotta, 1994), 548; and Alf Lüdtke, "'Helden der Arbeit' – Mühen beim Arbeiten. Zur missmutigen Loyalität von Industriearbeitern in der DDR," in *Sozialgeschichte der DDR*, eds. Hartmut Kaelble et al., 188.

4 Michael Beleites has shown that the roots of the peace, human rights, and independent environmental movement drew on debates in the West. See Beleites, "Die unabhängige Umweltbewegung in der DDR," in *Umweltschutz in der DDR. Analysen und Zeitzeugenberichte. Vol. 3: Beruflicher, ehrenamtlicher und freiwilliger Umweltschutz*, eds. Hermann Behrens and Jens Hoffmann (Munich: Oekom, 2007), 184–6. Sociologist Detlef Pollack is also convinced that for alternative political groups "the reception of Western critiques of capitalism and civilization [was] crucial." See Pollack, *Politischer Protest. Politische alternative Gruppen in der DDR* (Opladen: Leske & Budrich, 2000), 257. The thesis argued here about the importance of pioneers within East Germany itself is also discussed in: Astrid Mignon Kirchhof, "'Gelebte Überzeugung.' Das Wirken von Erna und Kurt Kretschmann für den DDR Umwelt- und Naturschutz," in *Agonie und Aufbruch. Das Ende der SED-Herrschaft und die Friedliche Revolution in Brandenburg*, eds. Jutta Braun and Peter Ulrich Weiß (Potsdam: Brandenburgische Landeszentrale für politische Bildung 2014), 190–211.

Kretschmanns – aged 74 and 76 years, respectively, in 1989 – were not part of it due to their age; however, the effect of the counterworld that they had created decades before became particularly potent during this period. On an intellectual level, the movement shows a convergence with the values that the Kretschmanns had been promoting for decades. Many environmental activists in East Germany – much like those active in the West – put the individual at the center of their thinking, envisioning how a person could seek a life in harmony with nature through a consumerism-critical lifestyle: in other words, they developed sustainable ideals. On a practical level, the East German environmental movement picked up the calls for creating nature conservation areas that the Kretschmanns had been suggesting since the 1950s, and introduced the national park program into the German Unification Treaty. Through this embedding in the structures of reunified Germany, the claims lost their counterworldly character and were integrated into nature conservation in the new Germany.

Holism, Life Reform, and Vegetarianism

Erna Kretschmann, neé Jahnke, was born in 1912 in Stettin; Kurt Kretschmann was born two years later in Berlin. Both of them had other careers before they dedicated their lives to nature conservation: Erna had been a kindergarten teacher and Kurt a tailor. We can trace the gradual development of their nature-oriented worldview in the 1930s and the influence of the Life Reform movement. Although the Kretschmanns are sometimes portrayed in the literature as pragmatists more than true "thinkers,"[5] this characterization is misleading; because the Kretschmanns stood for values that were always closely connected with everyday life, it sometimes

5 See Michael Succow, Lebrecht Jeschke, and Hans Dieter Knapp, eds., *Naturschutz in Deutschland: Rückblicke– Einblicke–Ausblicke* (Berlin: Ch. Links, 2012). The authors themselves call Erna and Kurt Kretschmann the "East German pioneers and trailblazers," 35–44, esp. 39–40.

creates the impression that they were more interested in concrete realities than in abstract matters of worldview. Nonetheless, they developed a clear set of ideas – drawing on the Life Reform movement as an inspirational model and combined with their own political convictions – that fundamentally shaped the way they led their lives. This worldview melded together a variety of ethical tenets and reform approaches: vegetarianism, critique of capitalism, abstinence from alcohol, rejection of technology and urbanization, pacifism, and a desire to connect with nature.[6] In an interview with Marion Schulz in 1990,[7] the couple described their paths to nature conservation. Kurt pointed to his critical attitude towards the city as a trigger for his yearning for nature; for Erna, it was meeting Kurt in the mid-1930s that changed her life.

A committed anti-fascist and pacifist,[8] Kurt Kretschmann resigned his position in a Berlin tailor shop in 1933 when the company switched its production to military uniforms because he "did not want under any circumstances to put his work in the service of the fascist state and its preparations for war."[9] Now unemployed, he spent much time hiking in the natural spaces surrounding Berlin. In the Brandenburg forests near Bernau, he became acquainted with Herbert Marquardt and subsequently took up residence with his new friend, living in a cabin, gardening, and beekeeping. Kretschmann adhered to a vegetarian diet and exercise regime

6 See Kurt Kretschmann, *Erinnerungen an meinen im Hitler-Krieg gefallenen Freund Herbert Marquardt* (Bad Freienwalde: Eigenverlag, 2002); as well as *Unsere Eß- und Trinkgewohnheiten unter die Lupe genommen. Erfahrungen aus 70-jähriger fleischloser Ernährung*, n.p., n.d., and *Gedichte gegen den Krieg*, n.p., n.d.

7 Marion Schulz, *Ein Leben in Harmonie. Kurt und Erna Kretschmann – für den Schutz und die Bewahrung der Natur* (Neuenhagen: Findling Verlag, 1999), 13. See also Daniel Fischer, "Zum Gedenken an Kurt Kretschmann, einem einzigartigen Menschen, der Naturschutzgeschichte," accessed August 22, 2017, <https://www.yumpu.com/de/document/view/7863164/zum-gedenken-an-kurt-kretschmann-haus-der-naturpflege>.

8 As an expression of their feelings about the war and the national socialist regime, the Kretschmanns named their son "Friedhart" (*Fried* meaning "peace" and *hart* "strong" in German), see Kretschmann, *Gedichte gegen den Krieg*, 63.

9 Kretschmann, Erinnerungen, 12, n 1.

and abstained from alcohol. The two men met Erna, who lived in the neighboring village Rüdnitz, and they both fell in love with her.

This period was also marked by exploration of various philosophical worldviews, some of which became important in the formation of the Kretschmanns' own life philosophy. Kurt later wrote:

> During this period, in the first summer, we [i.e. Kurt and Herbert] occupied ourselves intensively with the books that I hauled over. We wanted to drive forward our spiritual development and had discussions day and night about what we had read.[10]

The holistic thinking of the adherents of the Life Reform movement fostered their interest in non-European and particularly Asian traditions such as Buddhism and Confucianism. Among the thinkers that Kurt read and discussed with his friend Herbert Marquardt was Gandhi, whose principles Kurt would later write about in his poetry.[11] Mahatma Gandhi's concept of conscientious, non-violent resistance was inspired by Henry David Thoreau, who had lived in a cabin in the woods in Massachusetts for two and a half years in order, as he wrote, "to confront only the essential facts of life."[12] Thoreau's text "On the Duty of Civil Disobedience" became a foundation of the concept of resisting authority that Gandhi adopted in his own activities. Kurt and Herbert were particularly fascinated by Gandhi's apparent authenticity, the principle of non-violence that he preached and lived by, and his demand for political self-determination. Another source of fascination was Confucius' ideal of the "noble," the morally faultless person, who lives in harmony with the whole world. The philosophy of the Swiss Life Reformer Werner Zimmermann, whom Kurt had met at an event in Berlin in the 1930s, had the most enduring influence on Kurt and Erna Kretschmann. Zimmermann was an anti-capitalist advocate of society without control and of free socialism, as well as of Life Reform in the form of a new way of life, ecological agriculture, and free

10 Kretschmann, Erinnerungen, 21.

11 For example, in the poem "Mahatma Gandhi – 'Die große Seele,'" see Kretschmann, *Gedichte gegen den Krieg*, 65.

12 Henry David Thoreau, *Walden, Volume 1* (Boston, MA: Houghton and Mifflin, 1897), 143.

expression of sexuality.[13] Zimmermann's particular philosophy combined the desire for a complete renewal of the human way of life by returning to an original state of nature with the demand for an anarchistic form of economy called *Freiwirtschaft*. In 1934 Zimmermann founded the Swiss Economic Circle (*Wirtschaftsring*, abbreviated WIR – "we"), a self-help initiative that was simultaneously conceived as a "practical form of free socialism."[14] Connected with Zimmermann's model of living and economy was a demand for reforming sexual and dietary habits. He favored a new form of sexuality in which intercourse was decoupled from lust in the sense that orgasm should not be the center of the sexual act but love's consummation. As Kurt Kretschmann's philosophy of life slowly took shape through reflection and experimentation, this perspective led him to ask Erna whether she could imagine being intimate with his best friend Herbert: all three of them were bound together by an ardent friendship that they believed would enable the emergence of a sexuality built on these bonds of friendship without desire for possessions. The final step – coitus – was to be omitted. Erna agreed. This episode corresponds quite closely to Zimmermann's religious-ascetically tinged ideas of sexual friendship and "controlled lust" inspired by the karezza method.[15] Kretschmann later reflected on how they adopted Life Reform ideas in their lives:

13 Meike Sophie Baader, *Erziehung als Erlösung: Transformationen des Religiösen in der Reformpädagogik* (Weinheim: Beltz Juventa, 2005), 230–234. Also: Kretschmann, *Gedichte gegen den Krieg*, 64.

14 Günter Bartsch, *Die NWO-Bewegung Silvio Gesells: Geschichtlicher Grundriß 1891–1992/93* (Lütjenburg: Gauke, 1994), 27.

15 Originating in the United States, karezza is a form of sexual practice in which the usual culmination, orgasm, is avoided through self-control. It fit well into the German Life Reform movement's ambivalent attitude towards sex: on the one hand, the movement considered sex legitimate and the most natural thing in the world and was in favor of overcoming what were felt to be repressive sexual morals. On the other hand, the movement was dominated by a certain asceticism, a new form of enlightened prudery, and called for a sense of responsibility, restraint, and self-control. Karezza accommodated both, for it permitted sex while improving its value and avoiding the necessity for contraceptive measures. See XY, "Das Knospenwunder – Karezza, die Prophetie der unvollendeten Liebe," in *Propheten in deutscher Krise. Das Wunderbare oder Die Verzauberten*, ed. Rudolf Olden (Berlin: Rowohlt, 1932), 273–291; likewise:

> We refused to follow any fixed formula and lived with nature, had neither a calendar nor a clock, took our leisure whenever and as much as we wanted, without dividing the day into hours, the hours into minutes, the minutes into seconds. Eventually we were able to avoid all compulsions and worked when it gave us pleasure, went for a ramble when the desire called us, and slept when we felt it was needed.[16]

Like many of the leading figures of the diverse Life Reform movement,[17] Kurt and Erna Kretschmann also were committed vegetarians. The impulse for vegetarianism had its basis in a fundamental critique of the capitalist way of life, a reaction to the transformation of society through industrialization in the nineteenth century. Erna Kretschmann explained the link between the critique of capitalism and nature conservation as follows:

> I know people who need new furniture every few years. But if all of humankind follows this practice, more and more forests have to be cut down in order to produce the vast quantities of furniture. [...] Everything is oriented towards "throwing away" and "wearing out." We couldn't and wouldn't go along with that.[18]

And in fact, anti-capitalist reflections were also the basis for Kurt Kretschmann's path to vegetarianism. But while the majority of adherents

---

Ulrich Linse, "Sexualreform und Sexualberatung," in *Handbuch der deutschen Reformbewegungen*, ed. Krebs et al., 211–226 and Kretschmann, *Erinnerungen*, 82 f.

16 Kretschmann, *Erinnerungen*, 21.

17 Eva Barlösius, *Naturgemäße Lebensführung. Zur Geschichte der Lebensreform um die Jahrhundertwende.* (Frankfurt/Main: Campus, 1997), 217. See also: Heinrich Schipperges, "Zur Theorie der Lebensordnung und Praxis der Lebensführung in historischer Sicht," in *Lebensreform als ganzheitliche Daseinsgestaltung. 25 Jahre Eden Stiftung*, ed. Eden-Stiftung zur Förderung Naturnaher Lebenshaltung und Gesundheitspflege (Bad Soden/Taunus: Eden-Stiftung zur Förderung Naturnaher Lebenshaltung und Gesundheitspflege, 1988), 11–34. Likewise: Sabine Merta, *Wege und Irrwege zum modernen Schlankheitskult. Diätkost und Körperkultur als Suche nach neuen Lebensstilformen 1880–1930* (Stuttgart: Steiner, 2003), 93–97. Florentine Fritzen notes that vegetarianism originally meant a basically healthy dietary practice but not necessarily one without meat, as it was for Erna and Kurt Kretschmann for 70 years. See Florentine Fritzen, *Gesünder leben: Die Lebensreformbewegung im 20. Jahrhundert* (Stuttgart: Steiner, 2006), 41.

18 Marion Schulz, "Das Ehepaar mit den meisten Denkmälern," in *Ein Leben in Harmonie*, 43.

of vegetarianism were of middle-class background and were particularly afraid of becoming proletarianized and thus rejected any solution that involved any form of society radically different from the bourgeois one, Kurt Kretschmann identified with communist principles and continued to do so many years later, when he wrote about his vegetarian lifestyle:

> It was no coincidence that in "red Berlin" the KPD [Communist Party of Germany] was the strongest party. I, too, at 17, joined in enthusiastically as a sympathizer and marched along in all the demonstrations. [...] Then 1933: one thing after another happened. Hitler, financed with the millions provided by German heavy industry, came to power. I was distraught, but had previously pondered day and night how one might avoid capitalist exploitation. If I could succeed in reducing my expenditure for food – 14 marks a week – by half initially, and continue to reduce it, then it must surely be possible to shake off the yoke of the rightless and exploited proletariat. [...] It started in 1931. I wanted to determine whether one or two days of just fruit each week could have an influence on the amount of work to be accomplished.[19]

Just as it is today, vegetarianism was an approach to life that involved reflecting upon social and economic problems, such as the ecological consequences of excessive consumption of meat, and also had a moral component, such as respect for all living beings and a desire for a healthy lifestyle, a connection to nature, and a religious-spiritual life. Many nature conservationists in both West and East Germany believed that this desire for spirituality and contemplation could be met in nature that offered answers to ultimate questions. Motivated by this conviction, Kurt Kretschmann undertook repeated hiking trips as far as northern Italy as a young man, and later founded, together with his wife Erna, a hiking group in Bad Freienwalde that they led for 30 years.[20]

19 Kretschmann, *Unsere Eß- und Trinkgewohnheiten*, 8 et seq.

20 Noted by Kurt Kretschmann in the film "Kurt und Erna Kretschmann im Naturschutzmuseum," Studienarchiv Umweltgeschichte an der Hochschule Neubrandenburg [hereafter StUG], Bestand Kurt Kretschmann, 027–47, 2006. In addition to spiritual journeys, Kretschmanns wanderings were also an attempt to find a way out of Nazi-governed Germany, see Kretschmann, *Unsere Eß- und Trinkgewohnheiten*, 23.

Their experiments with alternative diet and Life Reform had to be halted with the onset of the National Socialist regime. In 1939 Kurt Kretschmann was conscripted into the military, released shortly thereafter because he was considered to be a danger to troop morale, but then made to work as a forced laborer in 1941 and later sent as a medical orderly with the Wehrmacht to the Eastern front. After he was sentenced to death for subversion of the war effort, he was able to desert in 1944 and hid for weeks in a pit underground in Bad Freienwalde, where Erna (now his wife) took care of him. The Kretschmanns experienced the end of the Nazi dictatorship as a complete liberation and wanted to express their gratitude towards the new government and participate in building and maintaining the new state. Kurt later wrote of this period: "Now we had to decide: what type of work will we take up? Vegetarianism was very important in our life course. We deliberated and then took up the cause of nature conservation, because through this we could reach more citizens."[21]

Frankfurt/Oder and Müritz: The Kretschmanns in the Early Years of East Germany

After World War II Erna and Kurt Kretschmann were members of the Freienwald chapter of the Communist Party; after the Socialist Party (SPD) and Communist Party (KPD) were merged to form the Socialist Unity Party (SED), they joined both the new party and the Cultural League (*Kulturbund zur demokratischen Erneuerung Deutschlands*), a state-run cultural organization. Erna Kretschmann was initially, until 1949, the district councilwoman for Public Education and developed the castle Bad Freienwalde into a culture center. In the same year Kurt Kretschmann became a nature conservation commissioner in the district Oberbarnim in eastern Brandenburg. Among his duties were surveying the area to identify

21 Kretschmann, *Gedichte gegen den Krieg*, 60.

– after the war – intact natural monuments and safeguard still existing nature and landscape conservation areas. Political goals, such as the land reform that started in 1945 in the immediate postwar period, were also reflected in the job description of the nature conservation commissioners. This person was responsible for the preservation of former manor estates, now transformed into people's parks. The commissioners remit also included planting hedges and stands of trees around the houses of the "new farmers," who were typically war refugees resettled on farms created by the land reform. Additionally, he or she (mostly he) was expected to recruit young people to the nature conservation effort, publish articles about their activities in the press, and give lectures. The creation of additional "Friends of Nature and Homeland" (*Natur- und Heimatfreunde*) chapters within the Cultural League was another duty.[22]

The Cultural League had been initiated by poet and politician Johannes R. Becher and authorized by the Soviet military administration in early summer 1945.[23] The initial goal of the organization was to mobilize intellectuals and artists to participate in a democratic new beginning after the end of the Nazi era. But the association's activities were also attractive to people in search of useful leisure time activities and social contact with like-minded individuals. Thus, the Cultural League was soon home not only to academics and artists, but also hiking enthusiasts, nature conservationists, and folkdance groups.[24] In November 1950 the Central Commission of Friends of Nature and Homeland was founded at a conference in Dresden and eight sections, later renamed "expert committees," were established. The Friends of

22 Brandenburgisches Landeshauptarchiv, Rep. 205 D, Nr. 1, Kurt Kretschmann, "Die Arbeit des Landesbeauftragten für Naturschutz, 14.10.1950." On the land reform, see Wolfgang Benz, *Auftrag Demokratie: Die Gründungsgeschichte der Bundesrepublik und die Entstehung der DDR 1945–1949* (Berlin: Metropol Verlag, 2009), 113 ff.

23 Bundesarchiv Berlin [hereafter BArch] DY 27, Nr. 841, "Letter to Mr. Bell, Information Section Control Branch, 6. 6. 1947." Sources differ regarding the precise date of the application and approval. See following letter BArch, DY 27, Nr. 841, "Brief of deputy military commander, Oberst Jelissarow to Johannes R. Becher, 25. 6 1945."

24 Ulrike Köpp, "Heimat DDR. Im Kulturbund zur demokratischen Erneuerung Deutschlands," in *Ethnografisches Arbeiten in Berlin: Wissenschaftsgeschichtliche Annäherung*, eds. Martina Krause et al. (Münster: LIT, 2003), 97.

Nature and Homeland expanded continuously within the Cultural League, to the degree that it became a problem for its leadership because in the 1950s "many local groups were so strongly represented that the Cultural League as an 'Organization of Intelligence' was scarcely recognizable as such."[25] Erna Kretschmann worked alongside these nature and homeland protection activists in the Central Expert Committee for Landscape Design and Nature Conservation and was also District Secretary for the Friends of Nature and Homeland in the Frankfurt/Oder chapter of the Cultural League.[26]

Nature conservation in East Germany rested on three main pillars: 1) government institutions, particularly the Ministry for Agriculture and Forestry (MLF); 2) the scientific consultation provided to the state by the Academy for Agricultural Sciences, and 3) societal involvement through the Cultural League.[27] In 1952 nature conservation was placed under the main department of forestry as the superior forestry office. The highest level nature conservation office was the MLF, followed by the forestry offices of the district councils, and finally those of the county councils at the lowest level. In 1951 Erna Kretschmann served as a nature conservation consultant for the Bad Freienwalde county council. In the same year, Kurt Kretschmann was appointed a special commissioner for matters of nature conservation by the MLF of the state of Brandenburg and headed the state office for nature conservation. During this period, he introduced a stylized owl symbol to designate nature conservation areas; after the enactment of the nature conservation act in 1954 it became the standard symbol across

25 Thomas Schaarschmidt, "Heimat in der Diktatur. Zur Relevanz regionaler Identifikation im Nationalsozialismus und in der frühen DDR," in *Zwischen Emotion und Kalkül: "Heimat" als Argument im Prozess der Moderne*, ed. Manfred Seifert (Leipzig: Uni-Verlag, 2010), 138. See also Jan Palmowski, *Inventing a Socialist Nation. Heimat and the Politics of Everyday Life in the GDR 1945–1990* (Cambridge: Cambridge University Press, 2009), 30.

26 Cf. curriculum vitae of Erna Kretschmann, in: Haus der Naturpflege e.V., *Erinnerungen an Erna Kretschmann (1912–2001)* (Bad Freienwalde: Eigenverlag, 2012), 18–19. Cf. Schaarschmidt, *Heimat in der Diktatur*, 138.

27 Hermann Behrens, "Naturschutz in der DDR," in *Wegmarken: Beiträge zur Geschichte des Naturschutzes*, ed. Stiftung Naturschutzgeschichte (Essen: Klartext, 2000), 206.

East Germany,[28] and in 1992 it was introduced in the federal states of former West Germany as well.

Kurt's activities in his role as special representative clearly show that he was determined to accomplish his nature conservation goals at any cost and accordingly things got messy occasionally. The following examples show that the Kretschmanns saw themselves at times as the experts in nature conservation and participated in turf battles with the authorities. In order to avoid administrative obstacles, Kurt advised other nature conservation commissioners to follow his lead and tell little white lies if necessary. In accordance with his own principles, anchored in the ideas of the Life Reform movement, that the individual should assume as much personal responsibility as possible, together with a socialist skepticism about private property, he regularly sent circulars such as the following, dated July 26, 1951, in which he offered instructions on how nature conservation could be effectively implemented in the face of opposition by private owners and how to overcome objections:

> In negotiations with private owners it is recommended, should resistance make itself felt, to explain that the district office for nature conservation has already decreed that the said object is to be designated as a nature conservation area, the state administrator and the mayor will receive news immediately and thus any alteration [in this plan] is subject to authorization. If, however, one acts according to the most laborious path dictated in the nature conservation act, it may happen that the property owner, in order to prevent the [object] being placed under protection, will reach for an ax and beat us to it.[29]

The Kretschmanns circumvented hierarchical structures in order to gain or maintain influence and to pursue their self-appointed role as apostles of nature conservation. In order to lend their demands greater

28 See "Kurt Kretschmann," in Hermann Behrens and Jens Hofmann, eds., *Naturschutzgeschichten: Lebenswege zwischen Ostseeküste und Erzgebirge* (Friedland: Steffen Verlag, 2013), 265.

29 BArch, DK 1. 3752, Bl. 181, Rundschreiben Nr. 4/51 des "Landesbeauftragten für Naturschutz des Landes Brandenburg," Kretschmann, betrifft "Richtlinien über die Erfassung von Naturdenkmalen," cited in: Hermann Behrens, "Wende-Wege," in *Ein Leben in Harmonie*, ed. Schulz, 94, n. 4.

efficacy, they addressed their petitions to not one, but many different government representatives simultaneously and thus created a certain audience for their concerns. Thus, they attempted to create democratic options for appeal that otherwise did not exist. Kurt described their approach as follows:

> Everyone thus knew that this letter had been sent to every possible agency and government office. This usually annoyed the people so much, that they accepted our suggestion and implemented it. Thus, of the 70 appeals at least 50 were realized through our efforts.[30]

Because there were initially no official established posts for nature conservation in the East German administration, the states[31] turned instead to special commissioners to manage nature conservation tasks. However, Kurt Kretschmann took on additional duties of his own accord until he had extended his scope of authority to that of an official state commissioner for nature conservation – something that had existed during the Nazi regime in all states and provinces of the German Reich since the passage of the Reich nature conservation act in 1935. He not only assumed the official title of the dismissed state commissioner of Brandenburg, Dr. Kurt Gruhl, but also likely hoped that he would thereby be granted the powers and authority that went with it. This shows that Kurt and Erna Kretschmann were not just victims of a misguided nature conservation policy of the State.[32] On the contrary, at times they were tiresomely zealous conservationists who extended their scope of influence beyond what was authorized. As expected, this led to ongoing conflicts with authorities: Together with his wife, he printed posters and brochures with the headers "State Commissioner

30 Schulz, *Ein Leben in Harmonie*, 30.

31 An administrative reform in 1952 essentially abolished the state-level government in favor of a stronger central government. A system of districts was implemented instead.

32 Kurt Kretschmann was an opinionated fighter for his cause and sometimes used wayward methods to push through his goals. See also Scott Moranda, *The People's Own Landscape: Nature, Tourism, and Dictatorship in East Germany* (Ann Arbor: University of Michigan Press, 2014), 61–62.

for Nature Conservation" and "Office for Nature Conservation of the State of Brandenburg"; this earned him a major reprimand from the central department of forestry in the MLF. After a series of clashes between Kretschmann and the MLF, he was eventually denied the right, listed in his contract, to call himself a "State Commissioner" of a "State Office for Nature Conservation."

Such violations of the rules brought the Kretschmanns repeatedly into conflict with state institutions, for their non-submissive, pugnacious behavior provoked the socialist state's need for control over its citizens, resulting in frictions with superiors and administrative offices. However, this did not lead to a loss of faith in the couple: Kurt Kretschmann's position as a special representative ended – in part due to the administrative reform, as a result of which district commissioners were appointed instead – however, he assumed a new position as a consultant for nature conservation at the Academy of Agricultural Sciences in Berlin. A year later the couple was entrusted with a new area of activity as heads of the Müritzhof, a remote farmstead in what was then the largest nature conservation area in East Germany and what is today the county Mecklenburgische Seenplatte. This was to become the site of the "Central Educational Center for Nature Conservation at Müritzhof," the first institution of its kind in the world.[33] The Kretschmanns led this pioneering institute for six years, from 1954 to 1960, and trained 1,200 participants in practical ways to implement nature conservation locally at the level of their villages and communities.[34] The Kretschmanns' duties at Müritzhof were diverse and included, for example, explaining the 1954 nature conservation act (which Kurt Kretschmann had been substantially involved in writing) and providing strategies for dealing with bureaucracies or overcoming resistance. Kurt Kretschmann noted:

33 Ludwig Bauer, "Naturschutzarbeit der 1950er und 1960er Jahre in der ehemaligen DDR," in *Natur im Sinn: Zeitzeugen im Naturschutz*, ed. Stiftung Naturschutzgeschichte (Essen: Klartext, 2001), 53.

34 Kurt Kretschmann, *Entstehung der Lehrstätte für Naturschutz "Müritzhof"* (Neustrelitz: Verlag Lenover, 1995).

> From the beginning it was our intention to bring all sorts of institutions to Müritzhof in order to have a wide-ranging impact. [...] The proportion of state officials will increase significantly. [...] In this manner our voice will be heard everywhere.[35]

During this period the Kretschmanns increasingly grappled with a dilemma: although they believed in the moral rightness of the socialist ideology upon which the East German state was based, that state did not give adequate attention to the matter dearest to their hearts: nature conservation. To be sure, unpaid involvement in the cause of nature conservation was generally supported by the state, which also passed a multitude of nature and environmental protection laws. However, these were by no means always acted upon, particularly if they were expensive or interfered with economic interests. The Kretschmanns' unconventional actions and their uncompromising insistence on changes being made in accordance with their expectations meant that they soon came up against bureaucratic limits. In 1960 the Kretschmanns ended their work at Müritzhof and made a radical decision: they moved back to Bad Freienwalde and opened a nature and cultural center, the "House of Nature Care," which they ran until 1982. There are various interpretations of the motives behind this decision: Gebhard Schultz, archivist at the Erna-und-Kurt-Kretschmann Archive, proposes that it was due to their ultimately "anarchistic" stance.[36] By contrast, Hermann Behrens, head of the "Study Archive Environmental History," states that the perpetual conflicts with state bureaucracies and nature conservation administrators were the reason for the couple's retreat.[37] Kurt Kretschmann himself indicates in a booklet on Müritzhof that the lack of state support was

35 BArch, DK 1, 20291, Bl. 135 f, cited in: Institut für Umweltgeschichte und Regionalentwicklung, ed., *Lexikon der Naturschutzbeauftragten. Vol. 3: Naturschutzgeschichte und Naturschutzbeauftragte in Berlin und Brandenburg* (Friedland: Steffen Verlag, 2010), 175.

36 See Erna and Kurt Kretschmann Archiv (herafter KreA), Online-Finding aid, accessed August 22, 2017, <http://www.haus-der-naturpflege.de/uploads/PDF/KreA_Online-Findbuch_2011.pdf>, 11. See also Behrens, "Wende-Wege," 106.

37 Behrens, 106.

the straw that broke the camel's back.[38] Probably it was a combination of multiple factors – lack of state support, exhaustion from the constant conflicts, a desire for freedom and escape from state control – that led the Kretschmanns to create their own nature space in Bad Freienwalde. Their Life Reform-inspired worldview doubtless played a significant role too: the Life Reform movement aspired not to reform of the state, but of the individual. The "House of Nature Care" was based on the basic idea that a lifestyle in accordance with nature can be carried out by individuals, but together the sum of these individual choices would result in the desired reforms – not, as in communism, a reform of production, but a reform of consumption like that advocated by the Life Reform movement.[39] Just as the Life Reform motto that declared "[Change] starts with yourself," the Kretschmanns turned to the individual to find solutions for problems of nature conservation. Kurt Kretschmann wrote: "Then we returned to Bad Freienwalde and began, entirely privately, to develop the 'House of Nature Care" on our own piece of land. We maintained it for 22 years without state support and continued to develop it."[40] The idea of an educational center that they had explored at Müritzhof was continued here and implemented in accordance with their own vision.

Counterworlds in the House of Nature Care: Development and Significance

Nature and environmental protection continued to be a central concern and field of activity for the Kretschmanns the rest of their lives. Their House of Nature Care in Bad Freienwalde fulfilled various purposes for them: it was the focal point of their activities, through which they

38 Kretschmann, *Entstehung der Lehrstätte für Naturschutz "Müritzhof,"* 16.

39 Cf. Barlösius, *Naturgemäße Lebensführung*, 170 f. and 198 ff.

40 Kretschmann, *Entstehung der Lehrstätte für Naturschutz "Müritzhof,"* 16.

inspired many visitors to become practitioners of nature conservation. But it also served as a haven that enabled them to dedicate themselves unconditionally to their cause, largely unaffected by external – that is, state – influences. At the same time, the Kretschmanns were able to successfully run the center and engage in dialogue with interested people from all over Europe precisely because the center remained associated with official structures and networks. The Kretschmanns' goal was always to continue their activities long-term and have an effect outside their immediate sphere of influence. Eventually their ideas were taken up by the environmental movement that gained momentum in the 1980s, and during German reunification, nature conservation areas – one of the Kretschmanns' key concerns – were included in the German Unification Treaty.

The development of a nature-conscious counterworld

According to sociologist Martina Löw, spaces emerge through action in a process she calls "spacing."[41] For the Kretschmanns, this space of action – the site of their emerging nature-oriented counterworld – was on the Boasberg in Bad Freienwalde. Here they built their private residence as well as additional buildings that housed the educational center.[42] They maintained a large garden that also served as a demonstration garden for visitors,[43] posted signs on the property featuring Kurt's owl symbol to indicate nature conservation areas, and gave symbolic names to the cabins:[44]

41 Martina Löw, *Raumsoziologie* (Frankfurt/Main: Suhrkamp, 2001), 158 ff.

42 The descriptions in this section of the buildings on the Boasberg property are from the webpage of the House of Nature Care, accessed August 1, 2017, <http://www.haus-der-naturpflege.de/index.php?page=bauwerke-im-haus-der-naturpflege>.

43 The demonstration garden is also described on the webpage of the House of Nature Care, accessed August 1, 2017, <http://haus-der-naturpflege.de/index.php?page=schaugarten>.

44 "Rooms are charged with symbolic force and thereby draw particular groups of actors to them" writes Cornelia Dlabaja in "Symbolische Markierungen im Stadtraum," in

one was called "Storchenhütte" [Stork Lodge],[45] reflecting their particular interest in this bird. In 1962 the Kretschmanns added a two-story guest house designed for four people; it was named in honor of Karl Kneschke, executive secretary of the Cultural League, who had offered tremendous support to Kurt during the development of the first East German nature conservation act in the 1950s.[46]

During the 1950s and 1960s additional buildings were erected on the property, with the Kretschmanns taking great care to ensure that the buildings conformed to their ecological standards and the requirements of their missionary approach to educational work.[47] With a floor space of only 35 square meters, their home was quite a small space to house four people;[48] thanks to custom-fit furnishings, effective use of the basement and the space under the roof, they succeeded in creating additional storage areas for firewood and for drying grain, as well as space for a workbench and an oven for baking bread.[49] Next to the house they installed an environmentally friendly, waterless toilet.[50] In one of the lecture buildings they were able to do without artificial, electric lighting by means of openings in the roof that let in sunlight.[51] Building costs, too, were kept to a minimum.

Positionen zur Urbanistik I, eds. Oliver Frey and Florian Koch (Vienna: LIT Verlag, 2011), 153.

45 See webpage House of Nature Care, section 4, accessed August 1, 2017, <http://www.haus-der-naturpflege.de/index.php?page=bauwerke-im-haus-der-naturpflege>.

46 See webpage House of Nature Care, section 3, accessed August 1, 2017, <http://www.haus-der-naturpflege.de/index.php?page=bauwerke-im-haus-der-naturpflege>.

47 Today their home has been made into a museum, accessed August 1, 2017, <http://www.haus-der-naturpflege.de/index.php?page=museum>.

48 Erna Kretschmann brought two children from her first marriage into the union with Kurt. See Haus der Naturpflege, *Erinnerungen an Erna, Anhang Biographie*. Their mutual child died in 1945 of diphtheria, see Kretschmann, *Gedichte gegen den Krieg*, p. 63.

49 See webpage House of Nature Care, section 2, accessed August 1, 2017, <http://www.haus-der-naturpflege.de/index.php?page=bauwerke-im-haus-der-naturpflege>.

50 <http://www.haus-der-naturpflege.de/index.php?page=bauwerke-im-haus-der-naturpflege>, section 2.

51 <http://www.haus-der-naturpflege.de/index.php?page=bauwerke-im-haus-der-naturpflege>, section 4.

An 8-meter-high tower called the "wobbling tower" was constructed for just under 100 marks and was one of many towers and observation points in Bad Freienwalde from which one could enjoy a panorama of the surrounding area.[52]

Through 1982 a lively flow of visitors lived and worked in these buildings, played music, ate meals together and through their discussions with the Kretschmanns became fellow campaigners for a nature-conscious lifestyle. The careful, deliberately staged design of the premises was meant to demonstrate to visitors that this was its own little world close to nature. The garden, the cabins with their various functions, the eco-toilets, ovens, use of natural light, and outlook towers showed that the Kretschmanns had created a space where nature was at the center. Because the Kretschmanns did not merely work on the Boasberg, but lived there year-round, they provided a real-life example of what a life in harmony with nature could look like. The "better life" realized in this counterworldly nature space that drew on models from the Life Reform movement in the nineteenth century was symbolically demarcated in the names of the huts and the placement of signs, thus binding it to a particular location, the House of Nature Care on the Boasberg.[53]

Perpetuation of the counterworldly space

For a space to continue to endure, it is not enough for it to be marked out physically and symbolically; it must also become institutionalized – that is, it must have a continuing effect beyond its own sphere of action and form other actions long term. And this, indeed, can be considered a fundamental goal of the Kretschmanns' efforts: their activities were intended to guide people to make long-term changes and act in harmony with nature. The Kretschmanns taught by example, living the lifestyle that they advocated to their visitors. For example, they experimented with solar energy thus

52 <http://www.haus-der-naturpflege.de/index.php?page=bauwerke-im-haus-der-naturpflege>, section 5.

53 Cf. Löw, *Raumsoziologie*, 170 and 204 ff.

showing that it was possible to reduce environmental pollution which was extremely high in the GDR through the use of brown coal.[54] The institutionalization of a space, however, requires that it is continuously the object of processes of perception, imagination, and memory. In other words, the Kretschmanns' counterworldly nature space could only exist as such so long as it was perceived and acknowledged by others.[55] And in fact, their contacts were many; over the years they presented their visions of a different society to their many visitors and created robust networks. In the 1970s, reports by informal collaborators with the East German Ministry for State Security (Stasi) suggested that more than 70,000 to 80,000 people had visited the House of Nature Care within the 15 years of its existence[56] – around 5,000 visitors per year.[57] The visitors attended training courses and lectures, and the Kretschmanns fed them with food from their own garden, which served as an educational demonstration for those interested in implementing their newly acquired knowledge in their own lives. Through these activities the couple gained friends and supporters of their ideas about living in harmony with nature. To be sure, not all of their many guests went on to become vegetarians or raise their own food, but many of them shared the fundamental idea that every individual could do their part to live a "better life." A poem by Kurt Kretschmann describes how he envisioned this:

> Whoever looks far beyond one's own self, will in the end conclude:
>
> Not materially – for that we had no desire
> We only wanted a better star.
> Where humans, nature and water and fire,
> Is not the crown of creation – a monster that is bizarre

54 Schulz, *Ein Leben in Harmonie*, pp. 25–30.

55 Löw, *Raumsoziologie*, pp. 158 ff.

56 Archiv des Bundesbeauftragten für die Unterlagen des Staatssicherheitsdienstes der ehemaligen Deutschen Demokratischen Republik [hereafter BStU], BVfS Frankfurt Oder, Abteilung X 300, Bl. 38.

57 Writer Gisela Heller mentions 7,000 visitors annually in her contribution to a pamphlet in memory of Erna Kretschmann. See, Heller in: *Erinnerungen an Erna*, 6.

> But where woman and man and child
> Are happy, joyful and reconciled.[58]

A space of action is always pervaded by social inequality in some form. Its particular constitution is thus always dependent on the possibilities available to the actors: knowledge, material goods, social position, and membership in associations, and thus social networks.[59] The Kretschmanns were not wealthy, nor did they have a position of high social status. East Germany was not a classless society;[60] the realm of nature conservation was highly revealing of this social stratification. Leading positions were occupied primarily by academics. Kurt Kretschmann was the only non-scholar to work on the creation of the 1954 nature conservation act. The Kretschmanns compensated for their lack of social status by their enormous amount of knowledge and the practical implementation of it, which gave them the reputation of leading authorities on the subject during their lifetime. Additionally, they were involved in the Cultural League and maintained this affiliation, and with it, an important source of contacts, even after they withdrew from government service to establish the House of Nature Care. These associations and the status that came with it allowed them to build networks that extended into the highest circles of society – in the GDR and abroad.

One such contact was to Professor Herbert Krauß. He was professor for physiotherapy at the Berliner Humboldt University and chief physician at the clinic for physical-dietary therapy in the municipal hospital Berlin-Buch, where he practiced naturopathy. Under his influence naturopathy was recognized as a clinical discipline of physiotherapy in the GDR.[61] Since a natural lifestyle on all levels was the base of Kurt and Erna Kretschmann's

58 Kurt Kretschmann, "Du bist für mich die richtige Frau," in *Erinnerungen an Erna*, 16; translated by the author.

59 See Löw, *Raumsoziologie*, 210 ff.

60 Heike Solga notes that social mobility in East Germany was often connected with loyalty to the system. See Heike Solga, "Systemloyalität als Bedingung sozialer Mobilität im Staatssozialismus am Beispiel der DDR," *Berliner Journal für Soziologie* 4 (1994), 523–542.

61 Claus Ruda, "Die Naturheilkunde in Berlin," *Berliner Ärzteblatt* 1999, accessed March 16, 2018, <http://www.claus-ruda.de/artikel/naturheilkunde.html#10>.

ideology, it was only a matter of time that the professor of physiotherapy and naturopathy and the experts of nature conservation got to know each other.[62] When environmental groups were founded in the GDR in the 1980s, they also had contact to Herbert Krauß. Informal collaborators of the Stasi reported that activists wanted to get in touch with physicians at the clinic in Buch to get information about naturopathy.[63] The Kretschmanns also maintained close personal contact to nature protectionists in the other German state and in other Eastern and Western European countries. One important contact was to Carl Duve, who was head of the Nature Protection Agency in Hamburg after World War II. He was also president of the National Park Association in Hamburg. When they met, Erna and Kurt Kretschmann discussed with him one of their lifelong ambitions, which was to create national parks in the GDR. Subsequently Duve suggested that the east bank of Lake Müritz could be integrated into a transnational European reserve. On the recommendation of Carl Duve, Kurt Kretschmann was to become the GDR representative in the International Union for Conservation of Nature and Natural Resources (IUCN). However, to the disappointment of Kretschmann, this did not eventuate, probably because at that time it was unclear whether East Germany could become a member in the international body. It took a further 10 years for the IUCN to decide that individuals and associations of the GDR could become members, but not the GDR itself as a state. This decision came too late for the couple because they had already withdrawn from most official positions to live a sustainable life within their self-created world of nature.

Contact with the counterworld created by the Kretschmanns often occurred by chance for example, when visitors active in the Cultural League or in the Friends of Nature and Homeland became acquainted with the couple. Others came to the House of Nature Care through informal routes, because they needed advice for planting their private garden or had heard about the Kretschmanns from someone else. Their were individuals visiting

62 Galina Iwanzowa, "Für immer in meinem Herzen," in *Erinnerungen an Erna Kretschmann (1912–2001)*, ed. Haus der Naturpflege e.V. (Bad Freienwalde: Haus der Naturpflege e.V., 2012), 10.

63 BStU, Akte "Reinhard Schumann," No. Berlin 161/91, Volume 5, 22.

them just as travel groups "to learn from Erna and Kurt something useful about nature."[64] Amongst their visitors were veterinarians, doctors, farmers, professors, musicians, writers, nature protectionists, teachers, politicians, students and very 'normal' citizens.[65] The Kretschmanns also spent considerable time corresponding with experts in their own country as well as in Eastern and Western Europe, and these contacts often extended to personal meetings and held over decades.[66] They carried out extensive research while preparing exhibitions, lectures, and publications. Their publications numbered "in the thousands."[67] Even before German unification (censored) films were made about the couple and their House of Nature Care, which in turn aided in the creation of networks with like-minded people and popular awareness of their ideas.[68]

Some visitors became acquainted with the Kretschmanns and their center as children who accompanied their parents on visits and continued to remain in contact as adults. One such person was Anke Jenssen, who later served as chair of the association House of Nature Care e. V. founded in 1993. Likewise, Michael Succow, later the Deputy Minister of Nature Conservation for East Germany, visited the Kretschmanns as a child. In the 1990s Kurt Kretschmann recalled:

64 See the presentation of the deputy president of the Association House of Nature Care, Sybille Knospe, on the occasion of a conference at the House of Nature in Potsdam on December 5, 2014: "Über die Mauer: Deutsch-deutsche Kontakte im Naturschutz," Online on the webpage of the House of Nature Care, accessed March 20, 2018, <http://www.haus-der-naturpflege.de/uploads/PDF/TagungPotsdam20141205_Rede_SKnospe.pdf>.

65 Cf. the contributions in the brochure *Erinnerungen an Erna Kretschmann*, ed. Haus der Naturpflege e. V.

66 KreA 279, letter from Erna and Kurt Kretschmann to Andrej Stollmann on January 15, 1981.

67 See Erna and Kurt Kretschmann archive online-Finding aid, accessed March 20, 2018, <http://www.haus-der-naturpflege.de/uploads/PDF/KreA_Online-Findbuch_2011.pdf>, Introduction of the finding aid, 13.

68 StUG, 16. 2011, Hartmut Sommerschuh, "22 Jahre OZON – Umweltberichterstattung in den Medien," 65–79, here 67 ff.

> I have never forgotten the day that 12-year-old Michael visited me the first time. He was sent by his astute mother in order to report on his ornithological observations in his home village. [...] When he later attended the extended secondary school in Bad Freienwalde, he often visited us in the evenings in order to study nature conservation literature. Thus my wife and I have accompanied him, on his development path until today, for 38 years. That he would become a fighter for nature conservation, we were certain. But we could not have guessed that the young Michael would one day be a driving force in this field, with brilliant farsightedness and creative power, [impelling] an upward trend in nature and environmental protection that [we] had never thought possible.[69]

The Kretschmanns' education of Michael Succow was not just limited to his early years; they also advised him later in life. Kurt Kretschmann warned him: "As an outside critic one cannot do anything useful for nature conservation."[70] Succow took this advice to heart early on. He got a doctorate in biology and received an appointment as professor at the East German Academy of Agricultural Sciences in 1984. Succow has stated on multiple occasions that the Kretschmanns were the spiritual foster parents of nature and environmental protection in East Germany and also of tremendous importance for his own life path.[71]

Wherever they went, Erna and Kurt Kretschmann passed on their philosophy of life to the people they met and thereby raised awareness of problems of nature and environmental protection that had received little attention in East Germany. With this awareness also came the development of countermeasures. The writer and editor Gisela Heller recalled:

> [Kurt Kretschmann] also drummed up the awareness that one cannot dominate nature, cannot exploit it in the long run with impunity, but rather one must recognize its laws in order to make it useful. For nature can survive without humans, but not humans without nature.[72]

69 KreA, 152, Kurt Kretschmann, "Prof. Dr. Michael Succow – aktiv und weitsichtig für den Natur- und Umweltschutz tätig," Aufsatz für den Freienwalder Kreiskalender, 7 pages, here 2–3, published after 1991.

70 Cited in Christian Siepmann, "Leben im Todesstreifen," accessed August 21, 2017, <http://nationale-naturlandschaften.com/index.php?q=leben-im-todesstreifen-das-tafelsilber-der-deutschen-einheit-wird-20>.

71 Most recently in the volume *Naturschutz in Deutschland*, 39–40.

72 Gisela Heller, "Geliebt und verstanden werden ist das höchste Glück" in *Erinnerungen an Erna*, ed. Haus der Naturpflege e. V., 6. It was also this maxim that Michael

Thus, through their contacts the Kretschmanns created a social arrangement that raised awareness and stood in opposition to the practical politics of the East German government. The nucleus of this counterworld, the House of Nature Care, constantly attracted the attention of individual actors and created shared perspectives on the environmental problems and categories of actions to be taken.

Many people in the Kretschmanns' circle of friends and guests remained true to the cause of nature conservation and have continued to work, even after Kurt and Erna's deaths, for influential organizations such as the NABU (*Naturschutzbund* [Nature and Biodiversity Conservation Union]).[73] Such people include Mechthild and Christoph Kaatz, who came into contact with the Kretschmanns as a result of their work founding the *Storchenhof* [stork ranch] in 1979, a rescue center for injured storks and other large birds in Lohburg (Saxony-Anhalt). The Kretschmanns were likewise active in the Cultural League for the protection of the white stork, publishing the bulletin of the "white stork working group";[74] and they were in contact with specialists abroad, both in the east and west. According to Mechthild Kaatz, these contacts still benefit the Storchenhof today: "it is also in their spirit that we continue the legacy of both the Kretschmanns in the NABU working group on white stork protection."[75]

For the supporters of the environmental movement that would later develop in East Germany, as well as for countless people who were involved only indirectly, Erna and Kurt Kretschmann were representatives and

---

Succow took to heart and adopted as the basic philosophy of the Michael Succow Foundation, founded in 1999 as the first non-profit nature conservation organization in former East Germany after unification. See <http://www.succow-stiftung.de/> accessed August 17, 2017.

73 The *NABU* is one of the biggest German non-governmental organizations that dedicates its work to conservation at home and abroad.

74 Some of the bulletins published by Kurt Kretschmann are in the holdings of the Bundesarchiv Berlin, see BArch, DY 27/9646. His 1978 proposal to the Central Committee for Ornithology and Bird Protection to found the white stork working group is in: BArch, DY 27/9721.

75 Mechtild Kaatz, "Ihr schaute die Menschlichkeit aus dem Gesicht," in *Erinnerungen an Erna*, ed. Haus der Naturpflege e. V., 12.

mediators of their nature-directed vision of a better life, but also early champions of nature conservation in East Germany, preaching their cause with a missionary fervor. When building their lifelong educational achievements, the two complemented each other well:[76]

> Much of what [Kurt] preached at the time with the tongues of angels, in fact acquired the force of law nationwide. Some who heard him then will object that Kurt appeared to them rather as an archangel who intervened with a flaming sword. Yes, he was a zealot, an uncomfortable figure.[77]

Erna Kretschmann, by contrast, seems to have been the calming influence both in their marriage and vis-à-vis their many visitors, who described her as a "wonderful comforter, a supporter, a quiet revolutionary,"[78] with great diplomatic skill[79] and a "levelling, but also energetic, goal-oriented manner."[80] Anita Tack, Minister of the Environment, Health, and Consumer Protection in Brandenburg from 2009 to 2014, described Erna Kretschmann's work for and within her counterworldly nature space as follows: "She sought like-minded people, brought them together and kept them together in view of shared values and goals."[81] It was precisely her gentleness and motherliness, her lived altruism and selflessness that secured, it seems, her place in the marriage, among the Kretschmanns' followers, and in German nature conservation more generally. Those around her, such as her spiritual foster son Michael Succow, describe her:

> Erna Kretschmann was considered the "mother" of East German nature conservation. All who met her were fascinated by this small, slight woman; her knowledge, her diplomatic skill, her critical confrontation with the time she lived in, her unselfishness, friendliness and kindness, her promotion of and action on behalf of nature conservation, her constant preparedness to stand by Kurt, even sometimes steer

76 Anita Tack, "Beeindruckendes Lebenswerk," in *Erinnerungen an Erna*, ed. Haus der Naturpflege e. V., inlay.
77 Heller, "Geliebt und verstanden werden," 6.
78 Heller, "Geliebt und verstanden werden," 7.
79 Tack, "Beeindruckendes Lebenswerk."
80 Mechthild Kaatz, "Ihr schaute die Menschlichkeit aus dem Gesicht," 12.
81 Tack, "Beeindruckendes Lebenswerk."

> [him]. She gave so many people impulses for their lives, was able to win them to the cause of nature conservation.[82]

The differences in Kurt and Erna's personalities were certainly an advantage to their missionary efforts at the House of Nature Care, because it enabled them to reach a wide variety of people.

State surveillance and the influence of the counterworldly space

In parallel with their activities at the House of Nature Care, Kurt and Erna Kretschmann continued to work to some degree within official structures and networks. Together with the Friends of Nature and Homeland, they not only worked to protect white storks, but also engaged in bat protection projects, marked out hiking paths and nature trails, and planned public green spaces. Their continued connection with the Cultural League – an organization solidly embedded in the system of the socialist state – is undoubtedly one of the reasons why they were able to run the House of Nature Care without interference from the state. As historian Thomas Lindenberger shows, East German citizens were, in fact, able to articulate interests and needs (and one might add: also to live and practice them), as long as serious conflicts with the authorities were avoided.[83] For the Kretschmanns, completely opting out of state structures would have provoked the authorities, as would promoting controversial ideas while working for state agencies or institutions; they succeeded in avoiding unwanted attention by being involved in topics the state had (increasingly) little interest in.[84]

82 Succow, *Naturschutz in Deutschland*, 40.

83 Thomas Lindenberger, "Herrschaft und Eigen-Sinn in der Diktatur. Das Alltagsleben der DDR und sein Platz in der Erinnerungskultur des vereinten Deutschlands," in *Aus Politik und Zeitgeschichte* (B 40/2000), 4, accessed May 15, 2017, <http://www.bpb.de/apuz/25409/herrschaft-und-eigen-sinn-in-der-diktatur?p=all>.

84 Katrin Passens, *Der Zugriff des SED-Herrschaftsapparates auf die Wohnviertel* (Berlin 2003), 22 ff.

At the same time, their nature center developed into a niche within East Germany[85] in which a certain scope for freedom was possible because the state regarded it as politically harmless. Them and other dissidents created niches in East Germany in order to escape "overpoliticization," but in doing so contributed to maintaining the stability of the system which has long been recognized by scholarship on East Germany.[86] What is unusual about the Kretschmanns' niche, or counterworld, is that the couple became pioneers of ideas that were later taken up in part by the GDR environmental movement in the 1980s. Their ideas caught on as soon as the conditions of the system changed in that general criticism of the socialist government was articulated publicly. Activist groups such as the movement New Forum (*Neues Forum*) and the citizens' initiative for establishing Müritz National Park, and scientists such as Michael Succow adopted the ideas promoted by the Kretschmanns.

However, this does not mean that the Stasi was unaware of the Kretschmanns' activities. Several decades of reports from informal collaborators make this abundantly clear. While Kurt Kretschmann was labeled as eccentric in the 1950s and his statements to the County Control Commission of the Party were treated as unreliable,[87] judgments about the couple became much milder over the next decades, varying from the opinion that they could "not evaluate political events"[88] to they had a "good attitude towards politics."[89] The Kretschmanns' accomplishments for nature conservation were even emphasized by some informal collaborators. On

85 Günter Gaus coined the term "niche society" (*Nischengesellschaft*) in the 1980s, see: Gaus, *Wo Deutschland liegt: Eine Ortsbestimmung* (Hamburg: Hoffmann und Campe, 1983), 156–233.

86 Janka Kuball, "Der Kulturbund Altlandsberg. Stadtpflege und Umweltschutz unter dem Dach einer staatlichen Organisation," in *Eigensinn in der DDR-Provinz, Vier Lokalstudien über Nonkonformität und Opposition*, eds. Kerstin Engelhardt u. Norbert Reichling (Schwalbach: Wochenschau Verlag, 2011), 20–67.

87 BStU, BV Frankfurt Oder, KD Bad Freienwalde, ZMA 4884, 15 and 23.

88 BStU, Bezirksverwaltung für Staatsicherheit (BVfS), Frankfurt Oder, Abteilung X 300, Bl. 40.

89 BStU, BV Frankfurt Oder, KD Bad Freienwalde, ZMA 4884, 3 and 13.

the whole, the Kretschmanns were portrayed as unpolitical people who considered nature conservation their most important aim.[90]

There are two main causes of the change in the Stasi's assessment: first, the East German state itself gradually changed from favoring a firm hand and strict implementation of communist rule under First Secretary of the Socialist Unity Party, Walter Ulbricht, to the policy of détente followed – at least for a time – under Erich Honecker.[91] Second, the Kretschmanns developed their sphere of influence in a remote, provincial area far from urban epicenters such as Berlin, Leipzig, or Jena, which were the focus of the authorities' attention. As recent research has shown, this tunnel vision was a mistake: regional centers played not an insignificant role in the erosion of the East German state.[92] The unspoken conclusion of the state, that the Kretschmanns did not have any political impact, must be seen as erroneous, at least in retrospect. For the counterworld that the Kretschmanns developed under the watch of state and Stasi had a remarkable influence in the 1980s. Michael Succow is one example of this, for in addition to his professional interest in environmental protection he was also involved in the nascent environmental movement. Together with Matthias Platzeck,[93] a civil rights activist and co-initiator of the *Grüne Liga* [Green League], he participated in the "Brodowin Dialogues," gatherings of a circle of environmentally concerned writers, scientists, and activists that were initiated by the writer and environmental and civil rights activist Reimar Gilsenbach.[94] When in 1989 environmentalists of the Cultural League called for the creation of a separate ministry for nature conservation, it was Succow who was at their head. Subsequently, after being appointed deputy minister of

90 BStU, BV Frankfurt Oder, KD Bad Freienwalde, ZMA 4884, 94.

91 Andreas Malycha and Peter Jochen Winters, *Die SED. Geschichte einer deutschen Partei* (Munich: Beck, 2009), 201 ff.

92 Cf. Braun und Weiß, *Agonie und Aufbruch*.

93 Platzek was part of the federal representatives' council of the Green League and took part at the Central Round Table of the GDR. For an autobiography of Matthias Platzeck see Matthias Platzek, *Zukunft braucht Herkunft: Deutsche Fragen, ostdeutsche Antworten* (Hamburg: Hoffmann und Campe, 2009).

94 Reimar Gilsenbach, *Wer im Gleichschritt marschiert, geht in die falsche Richtung: Ein biografisches Selbstbildnis* (Berlin: Westkreuz, 2004), 284 et seq.

the newly created Ministry for Nature and Environmental Protection and Water Management in January 1990, he became instrumental in establishing the first national parks, a lifelong dream of the Kretschmanns.[95] At the same time, the New Forum movement was also loudly calling for the establishment of a national park in the area around the lake Müritz – the oldest and largest state hunting region of East Germany. A citizens' initiative for Müritz National Park – headed by Succow – was formed to work out the details of the national park program and presented it to the Central Round Table in Berlin. The proposal was approved. Upon integration of East Germany into West Germany on October 3, 1990, one of the greatest and most ambitious nature conservation goals of the time was realized: the new Germany was five national parks, six biosphere reserves, and three nature parks richer. Included as part of the Unification Treaty, these protected areas are still considered the "family silver of German unification," as Klaus Töpfer, the West German Minister for Environmental Protection, phrased it at the time.[96] The legal framework for these conservation areas was the result of tenacious effort in what were literally the last days of a disappearing state.

Conclusion

The Kretschmanns adhered to the principle, inspired by the Life Reform movement, that individuals are capable of advancing changes in society, rather than relying on state reforms as a source of salvation. Based on this

95 The following discussion draws on Kirchhof, "Gelebte Überzeugung," 190 ff.

96 Ulrich Messner, "Nur einmal im Leben," *Nationalpark*, Nr. 149 (2010), 21–24; Succow, *Persönliche Erinnerungen*, 63–70; Arnulf Müller-Helmbrecht, "Als Westbeamter in den Osten," in *Naturschutz in Deutschland*, eds. Michael Succow et al., 71–79; Christian Siepmann, "Das Tafelsilber der deutschen Einheit," *Klartext* 19 (2009), accessed August 21, 2017, <http://klartext-magazin.de/47A/deutschland2/indexb93f.html?p=38>.

conviction, Erna and Kurt Kretschmann created a counterworld that stood in clear contradiction to the way of life promoted by the Socialist Unity Party. Counterworlds require a place in which their sphere of activity can unfold. In this case it was the property on the Boasberg in Bad Freienwalde. Here the Kretschmanns built facilities and created a nature space in which they showed by their own example what a life in harmony with nature can look like. This was never meant as a merely private refuge; on the contrary, they were missionaries of their cause to the many thousands of guests who visited it over the course of several decades. This resulted in a value transfer, and the Kretschmanns' followers implemented in their own spheres of activity things they had learned from the couple. The process, traced here, of developing and consolidating this counterworld clearly shows the difficulties that the Kretschmanns had to overcome: For example, by their own choice they had no access to positions in higher echelons of society. They chose not to have a source of income on principle, because they wanted to live "freely" and independently. The Kretschmanns met these basic needs by living nearly self-sufficiently from harvests from their land; at the same time, they built up social and political capital.[97] They gained the reputation of being preeminent authorities in their field; in addition, they remained Party members their entire lives and were involved with the Friends of Nature and Homeland in the Cultural League. Through their access to the association, new networks opened up, some of them even extending into the highest circles of society. The fact that they did not completely withdraw from official structures is surely one of the reasons that they were able to maintain the House of Nature Care without interference from the authorities. To stay involved was not only a strategic decision because they always agreed in principle with the moral basis and political goals of socialism.

With their nature center they created a counterworldly niche that was neither politically codified nor explicitly outside of the socialist state. Another contributing factor in the lack of interference was the fact that it

97 Pierre Bourdieu, "Politisches Kapital als Differenzierungsprinzip im Staatskapitalismus," in *Die Intellektuellen und die Macht*, ed. Irene Dölling (Hamburg: VSA Verlag, 1991), 33–39.

was located far from the urban centers that were the focus of state attention. Counterworldly spaces of activity were in fact possible in East Germany, if they were not in the spotlight of political attention and if they succeeded in remaining part of official structures in spite of their counterworldly quality and avoiding conflict with those in power. The Kretschmanns' niche was thus exceptional in that the ideals and worldview of their counterworld were adopted in the 1980s by parts of the environmental movement and integrated into the Unification Treaty. Their ideals caught fire when the conditions of the system changed to allow for criticism of the socialist regime to be articulated publically. The distinctiveness of East German environmental discourses and movements thus shows that environmental activism in the east was not just imported from the West but demonstrates the significance of intellectual precursors to the later environmental groups within East Germany. Considering that environmental data was declared classified by the East German government starting in the 1980s, it is clear that counterworldly structures can become a counteractive principle that works for enlightenment and against mechanisms that systematically ignore and suppress information.[98]

98 Andreas Dix and Rita Gudermann, "Naturschutz in der DDR: Idealisiert, ideologisiert, instrumentalisiert," in *Natur und Staat. Staatlicher Naturschutz in Deutschland 1906–2006*, eds. Hans-Werner Frohn and Friedemann Schmoll (Münster: Landwirtschaftsverlag, 2006), 594.

MICHEL DUPUY

Justifying Air Pollution in the GDR, 1949–1989

When the communists took power in East-Germany, they inherited an industrial state which had already seriously polluted the environment and damaged agricultural production.[1] Once the SBZ/GDR began rebuilding its industry, after the period of Soviet dismantling, in 1948, it reverted to its main energy source: lignite, also known as brown coal. In 1944, the area that would become the GDR had extracted 173 million tons of lignite. In 1946 this dipped to 108.4 million of tons, but by 1953 it had returned to 1944 levels.[2] Lignite is a dirtier-burning energy source than black coal, releasing far more sulfur dioxide (SO_2) and also dispersing a significant quantity of dust during its combustion. There was little the GDR could do about the problem, however. The construction of high chimneys to disperse pollution required time and money the GDR did not have, and until the beginning of the 1980s, there was no cost-efficient filter for SO_2 emission. Thus, from the beginning, air pollution was a problem at the heart of socialism; the authorities needed to either justify it to residents affected or remain silent. For the most part, they remained silent – newspapers acknowledged smog's presence in 1953, but then published nothing on the topic until December 1, 1989.[3]

Nonetheless, scientists in the GDR studied pollution extensively between 1949 and 1989.[4] In the 1950s, the focus was on dust, but after

1 See Gerhard Lenz, *Verlusterfahrung Landschaft* (Frankfurt/Main: Campus, 1999).

2 See Felix Christian Matthes, *Stromwirtschaft und deutsche Einheit* (Norderstedt: Libri, 2000).

3 "Nebel beeinträchtigt den Verkehr," *Berliner Zeitung*, December 10, 1953, 6; Martin Woldt, "Smog in unserer Hauptstadt," *Berliner Zeitung*, December 1, 1989, 8.

4 See Michel Dupuy, *Histoire de la pollution atmosphérique en Europe et en RDA au XXe siècle* (Paris: L'Harmattan, 2003).

the 1970s, SO_2 received significant attention. In the 1950s, pollution was local in its environmental effects, but it became regional at the end of the 1960s before crossing international borders with the advent of acid rain at the end of the 1970s. In the 1950s, authorities and scientists approached air pollution as only a technological challenge, but by the 1970s it acquired political and diplomatic relevance. All of these factors conditioned public discourse in the press, the literature, and the movies, but after the beginning of the 1970s, this public discourse was challenged from the outside (West German media) and from the inside with the emergence of an ecological movement within the Protestant Church, which helped to build local public space (as defined by Habermas) throughout the GDR.[5]

Thus, this chapter both concerns a public discussion of atmospheric pollution under the control of authorities and the emergence of a counter discourse shaped by factors inside and outside of the GDR.

From a Technological to a Political Question, 1949–1981

A technological question inside the socialist order, 1949–1971

In the October 11, 1953, edition of *Freiheit*, the organ of the SED (*Sozialistische Einheitspartei* [Socialist Unity Party]) in Halle, the journalist Paul Zänker lamented how "every vehicle, every gust of wind whirls up often whole clouds of flying ash and dust, which, as a result of the deficient street cleaning, float like drifting dunes along the roadsides."[6] On February 2, 1954, in the letters page of the newspaper *Neues Deutschland*, official paper of the Socialist Unity Party of Germany (SED), Marianne

5 See Ray Rühle, *Entstehung von politischer Öffentlichkeit in der DDR in den 1980er Jahren am Beispiel von Leipzig* (Münster: LIT, 2003).

6 Paul Zänker, "Flugasche, eine lästige Begleiterscheinung unserer Industriegebiete," *Freiheit*, October 11, 1953.

Schukay, an inhabitant of Berlin, wrote about the power plant of Klingenberg. She complained, "When we hang out our laundry to dry, it is soon unrecognizable because of the coal dust. Our windows and curtains are always black, also our balcony flowers and the fruit in the garden. In the interest of our health, please do something about it."[7] These complaints were renewed on February 28, 1954, again in the letters page. The newspaper responded on March 20, 1954.[8] Its editors assured readers that the power plant would be equipped with filters by 1955. In the meantime, the inhabitants had to limit their consumption of electricity during peak periods. Critical remarks continued, however. On October 12, 1957, in *Neues Deutschland*, during the annual congress of the hygiene section of the Academy of Sciences and the Society of Medicine in Leipzig, a journalist observed, "In Berlin, the power plant Klingenberg and the neighboring cement plant Rummelsburg discharge together every day large quantities of dust in the air. Another Berlin company blows per hour 'only' 45kg dust in the atmosphere. If a housewife shakes a duster out the window, she can count on a warning by the police, and a fine in case of recurrence. But what happens, when three Berlin companies 'shake' that much dust on our town each day?"[9] In July 1963, an East-German journalist again had to reassure readers, promising that "man will soon go through Bitterfeld again with open eyes" because the new power plant will be equipped in 1970/1 with an electro filter.[10]

In other words, pollution was not entirely ignored by East German authorities. Mentions of air pollution appeared in the inside pages of newspapers, sometimes ironically. Authorities insisted that a technological solution would emerge, but in the meantime, the authorities would make every effort to clear the dust from the streets. Even if pollution received attention, journalists could criticize the authorities only on their failures not on their politics.

7 "Ein Brief aus dem 'Mohrenland,'" *Neues Deutschland*, February 2, 1954, 6.
8 "Aussprache im Kraftwerk Klingenberg," *Neues Deutschland*, March 20, 1954, 6.
9 Dieter Hannes, "Es staubt und kracht zuviel bei uns," *Neues Deutschland*, October 12, 1957.
10 "Dem Bitterfelder Schmutz den Kampf angesagt," *Freiheit*, July 12, 1963.

Discussions of solutions had to link to an optimistic vision of socialist progress guaranteed through popular mobilization. This discourse looked forward to a better future because the scientific means to achieve this would come within the next few years. Instead of negativity about the state's failures, the discourse emphasized the responsibility of citizens to be far-sighted and ideologically sound. It also stressed the inevitability of short-term pollution problems to guarantee a higher standard of living. Thus, in November 1958, during the conference of chemistry in Leuna, the SED leader Walter Ulbricht launched the following concept: "*Chemistry* gives bread, prosperity and beauty!" Given the vital role of chemistry for improving living standards, the regime presented pollution as a necessary part of the economic development that must be tolerated. In an article from 1965 written by dipl.-Phil. Werr (press office of the chemical cooperative of Bitterfeld), the author recalled the importance of agriculture in the cooperative because of fertilizer and pesticide production.[11] On May 1, 1964, in the *Berliner Zeitung*, the writer Jutta Bartus reported what she experienced during three years with the workers of Leuna, as she prepared to write the screenplay of "Born under black sky": "The sky above me seems to be black as in other places. Fly ash stains the eyes – nevertheless I feel happy, incorporated within a large community of working people. They too fight against difficulties and opposition, but their will and their forces turn towards the future."[12] Finally, on July 2, 1974, the journalist Rudolf Eiter in the *Berliner Zeitung* wrote about the power plant of Klingenberg: "The chimney has to smoke. For the residents of a power plant, it is often unpleasant, but we need energy."[13] Different actors (journalists, writers, political leaders) endorsed the same discourse: building socialism shall be accompanied by a temporary acceptance of pollution by citizens.

In a similar vein, novelists naturalized pollution as an inevitable part of industrial growth and progress. Christa Wolf presented a description of a landscape at the beginning of her novel *Der Geteilte Himmel* [*They*

11 Phil Werr, "Sorgen mit der Bitterfelder Luft …," *Freiheit*, June 11, 1965.
12 "Seht nur – unsere Märchenwiese," *Berliner Zeitung*, May 1, 1964.
13 Rudolf Eiter, "Ein Werk verändert sein Gesicht," *Berliner Zeitung*, July 2, 1974, 8.

Divided the Sky], published in 1963. The novel began: "Like a clenched fist its smoky breath came driving up out of the hundred factory chimneys into the pure sky [...] this damned water that had reeked of chemicals for as long as they could remember – tasted bitter. But the earth still carried them and would do so as long as they were there."[14] At the same time, in another novel published in 1964 from Erik Neutsch, *Spur der Steine* [*Trace of Stones*], the narrator describes a grey town with dust, the river, here the white Elster (Leipzig) "is not whiter." The pollution is "naturally" present, as an element of the landscape, but these novels deal with other subjects few years after the building of the wall in Berlin. The division of Germany and the construction of socialism was for Christa Wolf a necessity, even if she provided for a sympathetic listening to the experience of workers.

Through the 1960s, media outlets did not falsify data on pollution in information available to citizens. For example, on February 13, 1967, in the *Freiheit* newspaper, the author explained that the cost of repairing damage to the neighboring forest, the Dübener Heide, was estimated at 40 million marks over the next 40 years.[15] The journalist wrote that a link existed between the statistics for bronchitis and air pollution. His source was official: The 16th plenary session of the council of the district (*Bezirk*) Halle. On July 24, 1967, on the science and technology television program *Umschau*, the following report was broadcast and entitled: "Smoke damages forests." The figures mentioned 200,000 hectares of damaged forest.[16] It was the result of a study of the forest center in Tharandt. At this time, there was no organization in the GDR which could relay this information and use it against the authorities. In the western bloc the environmental situation was not better, and there was always a solution. Concerning forests, the report focused on genetic research and the resistance of species to pollution because

14 Christa Wolf, *They Divided the Sky* (Ottawa: University of Ottawa Press, 2013), 3.

15 Dr. Tormyn, "EKB-Emissionskollektiv tilgt Flugasche," *Freiheit*, February 13, 1967.

16 "Rauchschadenforschung in Tharandt," *Umschau*, July 24, 1967. Accessed in *Deutsches Rundfunkarchiv* (DRA).

researcher attempted to adapt trees within a polluted environment.[17] Writers had the freedom to provide statistical data, but only if they showed that a socialist society was trying to find a technological resolution to the problem.

The "New Economic System of Planning and Management of the Economy" in 1963 attempted to introduce market control mechanism within a planned economy and influenced the ongoing discussion of pollution. In this atmosphere of reform, some environmentally alarmist titles appeared. For example, an article by Rudi Wetzel in the popular publication *Urania* in 1964 was titled: "threatened forests."[18] Again, in the same publication in 1966, the naturalist Reimar Gilsenbach issued: "Technology against nature."[19] In this article, the author denounced the madness of humankind in the exploitation of nature. Obviously, he recognized some efforts in the fight against dust emissions, but he underlined the lack of filter technology against SO_2. Photographic images also provided striking evidence of a pollution problem. In the 1964 article for *Urania*, Wetzel invited the reader to establish a link between the power plant and forest damage (see Figure 1). Normally, when the press spoke about pollution, it rarely included this kind of photography. Most of the images in the press represented a laboratory and some technological solutions. These articles reflected the spirit of reform, but they also continued to follow official expectations for optimistic and practical solutions. The authorities authorized these articles and did not have to criticize them as they contained these practical solutions. In other words, the environment was not a political issue even in these more alarmist pieces.

17 There was no technological solution to the problem of sulfur emissions from power plants which were economically feasible.

18 Rudi Wetzel, "Bedrohte Wälder," *Urania* 27 (1964): 657–661.

19 Reimar Gilsenbach, "Technik contra Natur," *Urania* 29 (1966): 12–21.

Figure 1: From article in science journal *Urania*. The caption reads: "2.5 Millions of tons of sulfur escape every year from the chimneys in our Republic. Through the exposure to sulfurous gases, forests in the Ore Mountains will soon die."[20]

20 Rudi Wetzel, "Bedrohte Wälder," 659.

Outside the pages of major journals and newspapers, the government kept informed of pollution problems, but it still demanded that information reach them through regulated channels rather than through public protest. The authorities, for example, listened to discontent through the *Eingabe* [citizen complaint]. Citizens also wrote directly to the companies involved (*Kombinate*) in order to obtain compensation. Thus, since 1965, some large companies had hired environmental protection officers (*Umweltschutzbeauftragter*) in order to respond to the increase in complaints.[21] In addition, authorities received information concerning pollution damage to forests and to the health of the inhabitants through scientists working within its own administration.[22]

The ideological turning point

In May 1971, Erich Honecker took power and decided to encourage consumption so as to pacify East German workers in a period where socialist citizens had manifested their discontent in Poland (December 1970).[23] At the same time, the environmental question became an increasingly important subject for government policy. In the GDR, on January 1, 1972, the SED created the Ministry of Environmental Protection and Water Management as a response to the 1972 United Nations Conference on the Human Environment in Stockholm. Eastern European countries (except Romania) boycotted the conference because the GDR could not participate.[24] Preparation for this event provided an opportunity to reflect on a more global approach to the environment and the use of resources.

21 Lenz, *Verlusterfahrung Landschaft*.

22 Dupuy, *Histoire de la pollution*.

23 Mark Landsman, *Dictatorship and Demand: The Politics of Consumerism in East Germany* (Cambridge, MA: Harvard University Press, 2005).

24 Michel Dupuy, "Science et pouvoir en RDA: la pollution atmosphérique transfrontière," in *Pouvoir(s) et environnement*, eds. Laurent Coumel et al. (Rennes: Presses universitaires de Rennes 2017).

Initially, socialist efforts to protect environment and public health received positive press. In Western media, particularly in France, Eastern European countries were portrayed as models of environmental protection, with the Baïkal being a key example.[25] Radio Free Europe made a broadcast story on June 5, 1972, during the Stockholm's conference: "Environmental Problems in the GDR." It discussed the efforts towards a reduction in water and air pollution as well as noise pollution in the GDR and specified the laws and their applications in the relationship between companies and local authorities.[26]

By the end of the 1970s, however, the tone of western reports became harsher, and there was more criticism of the GDR.[27] In this context, the GDR now had to defend itself against criticism which came from the West. Here the goal was to convince television audiences that in the FRG the environmental situation was worse than in GDR. On the defensive, the GDR with its weekly television program *Schwarzer Kanal* [Black channel] denounced the West's lack of respect for the environment, based on capitalism's drive for profit. The GDR commentator admitted "Also, we have some rivers, which are not good for swimming, and air which is not perfume." But he also insisted, "We are on the right path with the right measures and effective means (and proven success), to cope with these side effects of industrialization."[28]

The East German debate about pollution was further complicated by Honecker's embrace of socialist consumerism. East German philosophers such as Wolfgang Harich, *Kommunismus ohne Wachstum? Babeuf und der 'Club of Rome'* [*Communism without growth? Babeuf and the 'Club of Rome'*]

25 Michel Dupuy, "The French media and the environmental problems in the communist countries during the Cold War" (Paper presented at the conference "Not just Chernobyl. Ecological Dimensions of the History of Communism," Poznan, Poland, April 21–22, 2016).

26 "Environmental Problems in the GDR," *Radio Free Europe*, June 5, 1972, accessed August 8, 2017, <http://osaarchivum.org/files/holdings/300/8/3/text/26-2-119.shtml>.

27 "Umwelt: Lenkt die DDR ein?," *Der Spiegel* 39, September 25, 1978, 18–19.

28 Karl-Eduard Schnitzler v. "Umweltverchmutzung," *Der Schwarze Kanal*, December 10, 1973, accessed August 8, 2017, <http://sk.dra.de/grape/seite6.htm>.

(1975), and Rudolf Bahro, *Die Alternative* (1977), published their reflections in West Germany. These writers, through a Marxist reading, expressed their opposition to socialism consumerism, because communism did not mean mollifying the population with consumer goods as was the practice in capitalist countries.[29] Harich wanted to halt car manufacturing in the GDR because it encouraged private property. Finally, he wanted to stop economic growth in Eastern Europe in order to solve the "ecological crisis."

These tensions reached a climax with the publication in 1981 by Fischer Press in West Germany of the novel Monika Maron's *Flugasche* [*Fly Ash*], which described a journalist's visit to the town of B. (Bitterfeld), "the most polluted town in Europe."[30] This novel was proposed in May of 1976 in the GDR with the following title: "And tomorrow I come again." However, literary authorities wanted "a book that corresponds to our artistic and ideological principles." After lengthy negotiations, Maron refused to accept more corrections and sent her book to the West.[31] In other words, Maron refused to accept the principle that all discussion of pollution retain optimism in technology and socialism's unique ability to solve these problems.

This politicization of the environment and the criticism from western media on the environmental damage in the GDR led the authorities to increase their control of publications and documentaries on this theme. The goal was ideological: to show that the socialist states were on the right path, but had to also overcome the inheritance of capitalism.[32] These prerogatives were not new, but authorities became increasingly defensive and aggressive in pressing authors to follow these guidelines.

The growing pressures on writers can be seen in the negotiations over a series of four documentaries called "Nature and Man" broadcast in October of 1974. The original plan dates from November 1971. Its creators initially presented a document to the Ministry of the Environment and

29 See Alexander Amberger, *Bahro – Harich – Havemann: Marxistische Systemkritik und politische Utopie in der DDR*. (Paderborn: Schöningh, 2014).

30 Monika Maron, *Flugasche* (Frankfurt/Main: Fischer, 1981).

31 BArch DR 1 16910, Büro für Urheberrechte, "Aktennotiz," January 19, 1981.

32 For example of a documentary doing this, see *Deutsches Rundfunkarchiv* (DRA), "Natur und Mensch."

the Ministry of Sciences and Technology on November 16, 1972, with the following title: "With both legs on the Earth." The director was obliged to modify the content because it did not align with the official policy of the 8th congress of the party, and the image of the GDR presented in the documentary seemed isolated in respect to other socialist states. A new script was proposed on November 20 but then refused. The final script was accepted on March 30, 1973. In this documentary, damage to the environment in Western countries was highlighted. Discussion of damage in socialist countries was accompanied by practical solutions and by citations from Engels. Thus, for the town of Bitterfeld, it was stated that 300 tons of ash fell every day on this town, but that a new power plant would enter into service in 1975 equipped with an electric filtering system. The purpose of nature was to serve human beings. Forests produced timber and were a place to reoxygenate workers through its recreation areas.

In the publishing world, the publication of a book about the environmental issue from Horst Paucke and Adolf Bauer, both members of the Academy of Sciences, was equally emblematic of this ideological control of works on environmental issues. A first version is presented in September 1976 with the following title: *Environment – Ideology – Reality*. In 1978, this book was already submitted for review, yet a rapporteur wrote: "The problems are more acute, a wealth of alarming information is in the media."[33] Despite the importance of this book, or perhaps because of it, the rapporteurs, most of them from the social or economic sciences but all authorities in their field, reinforced the ideological content, insisting on the role of the party and the solutions found within a socialist society. In February 1979, some changes had to be made. The title became: *Environment – Problems – Solutions*. In March 1979, the authors received authorization to publish with the following title: *Environmental problem challenge to humanity*.[34] These negotiations relating to content show a certain resistance from the authors. Paucke and Bauer wished to "ecologize" society with recycling and a higher productivity to liberate leisure time for the worker.

33 BArch SAPMO DY 30 21775, "Umwelt – Ideologie – Realität," February 10, 1978.

34 Horst Paucke and Adolf Bauer, *Umweltprobleme Herausforderung der Menscheit* (Berlin: Dietz, 1979).

The newspapers continued to operate in the same manner: a problem, a technological solution and always reassuring the public that pollution was measured continuously (*ständig*). For example, an article in *Das Volk*, on January 15, 1976, was particularly significant. It was an interview from Dr. Bernhard Schulze, director of the environmental hygiene institute in Erfurt. In the article, the tone is a bit alarmist: "Our environment is no longer paradisiac. The continuously increasing industrialization has its drawbacks. The waste products of progress can become a danger for our health," However, Schulze also reassured readers. Some measures had been taken, but he emphasized the impossibility of solving all the problems at once: "this costs time and often a great deal of money." Later, when asked about a possible consequence of smog in Erfurt, Schulze deferred answering the question and concluded: "We must not allow this confusion as a result of western fear in the population."[35]

Controlling the discourse on environmental data

At the end of the 1970s, the context changed again. Beginning in 1978, the Ecclesiastical Research Center (*Kirchliches Forschungsheim*) in Wittenberg became one of the centers of the nascent environmental movement in East Germany. Sparked by the UN's 1972 Stockholm Declaration, which laid out a vision for ecological sustainability alongside fundamental human rights, the Wittenberg Center began researching and discussing the environment, and in 1980 published seminal booklet, entitled "The Earth Needs to be Saved: Environmental Crisis, Christian Faith, Options for Action."[36] The Center also held an exhibition that probed the relationship between "Man and Nature."[37] Through publications and exhibits, the Center helped spur the formation of several grassroots ecological groups

35 "So schnell geht uns die Luft nicht aus," *Das Volk*, January 15, 1976.

36 Hans-Peter Gensichen, *Die Erde ist zu retten: Umweltkrise, christlicher Glaube, Handlungsmöglichkeiten* (Wittenberg: Kirchliches Forschungheim, 1982).

37 See Hans-Joachim Veen, et al., *Lexikon Opposition und Widerstand in der SED-Diktatur* (Berlin, Munich: Propyläen, 2000).

interested in exploring the root causes of environmental problems, which they believed were a result of the development of an industrialized society itself. They were not interested in simply accepting the standard Party line of the SED state any longer.

This questioning of industrial and technological progress spread beyond Wittenberg to Rostock, where the staff writers at the Honstorff publishing concern challenged the East German Minister of Environment, Hans Reichelt, to a debate on the environment and the veracity of the information being released to the public about it. Held at the Academy of Art in Berlin in November 1981, participants included Andreas Zecher, who openly claimed that the population was not really being informed on how acute pollution was, and Dorothea Kleine, who worried aloud that "The increasing lignite production is celebrated as a victory. But I feel threatened by a polluted environment."[38]

This concern was accompanied by an increase in visible damage to the forest. Thus, a report by the Stasi in April of 1981 estimated that 300,000 hectares of forest were damaged by pollution and that 45 percent of the population lived in an air-polluted area. Above all, the number of *Eingaben* [complaint letters sent to the Party or State] increased, notably with some from the church. It was a sign that the population did not trust the information from the authorities and the media.[39]

The Stasi analyzed the publications from the West, which increasingly presented a negative image of the GDR and, above all, of the socialist system. In their western publications, their sources included data published in the East-German press and in scientific reviews. In other words, the Stasi believed that the GDR provided serious ammunition to the West for criticism. Finally, the Stasi argued that East German authors did not emphasize enough the difference with the capitalist countries, based on the quest for maximum profit. The Stasi complained

38 BArch DK 5 188, "Niederschrift zum Gespräch des Stellvertreters des Vorsitzenden des Ministerrates und Ministers für Umweltschutz und Wasserwirtschaft, Dr. Reichelt, mit Schriftstellern und Mitarbeitern des Hinstorff Verlages am 19. November 1981, 15.00 bis 18.15 Uhr, Akademie der Künste."

39 BStU MfS, Sekr. Mittig, Nr. 19276.

that the "scientists, creative forces and members of the intelligentsia in the GDR diffuse pessimism and skepticism." In other words, they believed that the press had to be mobilized in cooperation with scientists on an ideological level.

In this Stasi report, the author called for a censorship of data in the press and for scientific reviews – although it should be noted that as early as March 19, 1974, the regime had begun censoring the publication of new environmental data. On November 16, 1982, the crucial decision was made by the authorities to reclassify all data on pollution as a state secret. This prohibition occurred just a few months after the GDR ratified the 1979 Convention on Transboundary Air Pollution, brokered through the United Nations Economic Commission for Europe (UNECE), which publicly bound the GDR to lowering air pollution in an international context, making air pollution data a matter of international relations and thus security. Thus, an external and an internal pressure resulted in the censorship of environmental data. The GDR was in a defensive position and increasingly silenced any discussion of pollution. There was no more public debate or discussion of environmental data after 1982.

A Contested Discourse in a Fragmented Public Place, 1982–1989

Rushing to a dead end

Censorship and ideology combined to undermine the credibility of the GDR, but the problem was fundamentally one of increasing pollution. And one reason that pollution increased in the GDR had to do with the USSR itself. Since 1975, the USSR had gradually adjusted its prices to that of the world market, and so it had to cut back on the amount of oil it delivered to the GDR. In 1981, the USSR cut back its delivery of petroleum to the GDR by 9 percent, causing a shortfall of 2 million tons, which severely impacted the synthetics and energy industries.[40] This meant that the GDR had to increase its usage of brown coal to make up for the loss of petroleum, and so between 1980 and 1985, lignite extraction in the GDR rose from 258 million tons 312 million tons. With more lignite in usage, SO_2 emissions rose from 4,264 kt in 1980 to a peak of 5,560 kt in 1987 (they would subsequently decline to 5,209 kt in 1988).[41] This was a problem for the GDR because, in 1985 in Helsinki, it had ratified and extension of the 1979 UNECE Convention on Transboundary Pollution, entitled the "*Protocol* on Long-Range Transboundary Air Pollution on the Reduction of Sulfur Emissions or their Transboundary Fluxes by at least 30 per cent." This meant that the GDR was bound to reduce its SO_2 emissions by 30 percent, at the very moment it was forced to rely on increased lignite usage.

Furthermore, the public impression of the damage caused by SO_2 had increased dramatically as well, especially in the West, where the concern over forest death (*Waldsterben*) had risen to near crisis levels. The concern extended to the East German forests, including in the area of the Erzgebirge [Ore Mountains], important symbolically and economically both on a regional and national level. Since 1982, the western media illustrated the forest damage caused by pollution with in images and documentaries, often using the Erzgebirge as their focal point. In part this was because the inhabitants of the Erzgebirge began complaining to GDR authorities about air pollution and its deleterious impact on the ecosystem in the Erzgebirge. A report from the Ministry for the Environment and Water Management to the Secretariat of the Central Committee remarked that in one 18 month period in from 1979–80, 220 complaints were received by the Ministry from citizens concerning the air pollution in the Erzgebirge.[42] As the *Waldsterben* issue heated up in the West and became a more frequent topic of discussion in the West German media, including RIAS (the West Berlin radio station that was frequently listened to by East Germans),

40 Eberhard Kuhrt et al., *Die wirtschaftliche und ökologische Situation der DDR in den 80er Jahren* (Opladen: Leske & Budrich, 1996).

41 Kuhrt et al., *Die wirtschaftliche und ökologische Situation*, 227.

42 BArch SAPMO DY 30 1738, "Bericht über die Luftverunreinigungen im Gebiet des Oberen Erzgebirges," November 1980.

more East Germans began to connect the issue of air pollution to dying trees, when they observed dying trees. Not only that, but when people put these observations together with the fact that their state was still officially blaming other causes, such as abnormally cold winter temperatures, for any damaged or dying trees they saw, they began to believe (or became more convinced) that their government was the one lying to them. The complaints to the Environmental Ministry began climbing, from 200 in 1980 to 463 in 1982, and eventually 676 by 1988.[43] Soon, East Germans were writing to *West* German radio, such as an East Berliner who wrote to RIAS in 1986 explaining that she was in the Erzgebirge two weeks ago, stating that she was "shocked," because there was no more forest and "just a few years ago, our friends [in Erzgebirge] didn't want to believe, that the forest was damaged and blamed the damage on the cold winter weather."[44]

This letter was read on the air, contributing to a feedback loop between East Germans and the West German media, which led to increased coverage in the West of East German pollution in the western media, which found that there was an eager audience for coverage of East German pollution both in the West *and* in the East, though likely for different reasons. So for example, the Sunday morning agriculture program on RIAS steadily began increasing its coverage of environmental issues after 1983, incorporating a frequent segment in which scientists would read papers containing research on environmental science. East Germans could hear the station, and those who could not get a signal could obtain copies of the transcripts through West German intermediaries.[45] There were other means of transmitting information about the environment from the West to the East, as well; beginning in August 1987, a West Berlin station called "Radio Glastnost," which was established by environmentalists, also began broadcasting information about the East German ecology into the GDR.[46]

On the other side, it was the GDR that held itself up to the *West* as a model of ecological conservation. At the height of the *Waldsterben* crisis, in

43 BStU MfS, Sekr. Mittig, Nr. 235.

44 DRA Hist Archiv 02/92/56, *Hörerpost* 08/1984/05/1985.

45 DRA Hist Archiv 02/92.

46 Paul Hockenos, *Berlin Calling* (New York: New Press, 2017).

which some scientists had predicted the complete extinction of the German forest within five to 10 years,[47] the GDR invited several West German politicians to tour the GDR and learn about East German techniques for remediating air pollution between 1983 and 1987, including the Bavarian Minister of Development and Environment, a delegation of Green Party members, a representative from the SPD, and the Minister of Labor and Social Affairs from Rhineland Palatinate.

These visits were highly organized, beginning with an initial discussion with GDR Minister of Environmental Protection Hans Reichelt, followed by an excursion to the Dübener Heide, the forest downwind of Bitterfeld which had been damaged by the Bitterfeld plant emissions and where pollution-resistant species of flora were being introduced. It ended with a visit to the experimental forestry station in Tharandt, where these pollution-resistant species were developed and produced.[48] Guests were also shown the power plant in Vockerode, east of Dessau and north of Bitterfeld, where a process invented in East Germany involving the addition of limestone succeeded in trapping 40 percent of the sulfur emissions.[49] The tour carefully avoided the Erzgebirge and Bitterfeld itself, which were polluted, but more importantly, which were already well known in the western media as place names synonymous with East German pollution.

In addition, the GDR attempted to portray itself as the victim of West German pollution, in part because East Germany was situated most downwind of West Germany, so that, especially near the border, emissions from West German facilities ended up in East Germany. In 1983, for example, the East German press reported about West German plans to construct

47 Michel Dupuy, "La Dübener Heide: un massif forestier entre enjeux scientifiques et politiques, 1957–89," in *Débordements industriels. Environnement, territoire et conflit XVIIIe-XXIe siècle*, eds. Thomas Le Roux and Michel Letté (Rennes: Presses Universitaires de Rennes, 2013), 247–267.

48 BArch SAPMO DK 5 1514, "Besuch einer Delegation der Fraktion der Grünen im Bundestag der BRD vom 1.-5.9.1986 in der DDR, ZK-Vorlage."

49 This process was piloted by the Institut of Energetic in Leipzig in 1985. It was the result of a program launched in 1967: W. Kluege, "Konzeption zur Verringerung schädlicher Auswirkungen der Energie- und Kohlebetriebe auf die Umgebung Teilkonzeption: Kraftwerke," *IFE-Bericht Nr. 2/1763/67 F* (1967).

a power plant in the town of Buschhaus with a 300-meter-high chimney, and no desulfurization equipment. The implication was that one could save money on desulfurization by building on the border and letting the wind take the pollution into East Germany.

But the East German media left out the damage being done to East German forests throughout the republic by East German emissions, only referring to air pollution in terms of air quality in cities, mainly, smog. When forests were discussed in the media, the conversation focused on the economic uses of the forest, saying that "the forest is a valuable raw material, source of the life, of health, friend of mankind" and claiming that, to the extent that there was damage, this damage came from storms, insect and frost damage, not the socialist state.[50] In fact, the use of the word "*Waldsterben*" was forbidden, and the Stasi closely watched over the use of photographs of trees damaged by pollution.[51]

When the press did speak about local measures to reduce dust or SO_2 emissions, it was always in general terms of "success" without using any actual data or figures, and therefore East Germans could not put the claims, even if they were technically true, into context. One example of this was the Vockerode limestone process. Beginning in 1984, the limestone additive was touted in the media and by GDR politicians as a great success in combating air pollution, and it was – it trapped 40 percent, and sometimes upwards of 60 percent of SO_2 emissions. What was left unsaid, however, was that this process itself was highly inefficient, because Vockerode was one of the smaller power plants in the GDR, and for the larger power plants to be able to utilize the same process, unsustainably large quantities of limestone would be needed. An estimate for one of the larger power plants in the GDR, Boxberg, estimated that in order to achieve the same results as Vockerode, 10,000 tons of limestone would have to be delivered every day.[52] Not only that, but the fact was that Vockerode was near to a

50 Rudolf Rüthnick, "Unser Wald – Rohstoffquelle und Stätte der Erholung," *Einheit*, November 1984.

51 Interview with Hans-Günther Dässler, Tharandt, August 17, 1999.

52 BArch DK 5 2133, Guido Thoms, "Beschluss über die Konzeption für den Einsatz des Kalkstein-Additiv-Verfahrens in der DDR," October 1983.

limestone quarry in Thalheim; many, if not most, power plants did not sit so close to a source of limestone. Except for one article in the *Wochenpost* in 1986, the press ignored these factors, and only focused on the success of the Vockerode process.[53]

This kind of obfuscation clashed with the fact that the East German public could access West German media and its reporting on pollution in the East – whether by TV, radio, or other means. The cognitive dissonance that this created contributed to making East Germans skeptical of the information that was being put out by their government.

Doubt and resistance

There were some instances of the East German press attempting to address the problem in subtle ways, despite the fact that the topic was generally taboo. For example, in a 1984 interview with Rolf Opitz, President of the Leipzig City Council, the *Leipziger Volkszeitung* related some of the concerns that its readership had about the environment in and around the city, writing: "Many of our readers write that it remains their wish to have a beautiful environment again." To the first remark, Opitz replied with the standard Party line, making reference to the inheritance of capitalism and about the efforts being made to overcome its legacy, maintaining that air pollution had declined by 10 percent since 1976 (though again, without giving any data to provide crucial context).[54] Another question the newspaper related from readers was: "We keep hearing that some existing plants don't use all of their desulfurization capacity and switch off their filter during the night?" In response to this, Opitz confirmed that the plants were being inspected by the hygiene inspection department of Leipzig and "switching off the filter for egoistic operational reasons was a thing of the past" because of the penalties that would be imposed. The *Volkszeitung* kept after the issue of filters not working or not being used

53 Ina Pachmann, "Ein Prototyp überraschte angenehm," *Wochenpost*, November 2, 1986.

54 "Die 'Dicke' Luft wird Jahr für Jahr sauber," *Leipziger Volkszeitung*, January 11, 1984.

properly, finally getting a member of the Leipzig Council for Environmental Protection and Water Management, Dieter Halbig, to admit that: "In the 55 cases where specified [pollution] limits have been exceeded, it was because filter installation wasn't carried out on schedule."[55] He noted that the companies in question were sanctioned, but he did admit that there had been some failures.

These kinds of partial admissions from low level officials were not enough to assuage the concerns of ordinary East Germans, some of whom used public meetings at local level to openly express doubts about the information they were being given, which caused difficulties for the authorities, both state and party. Thus, in September 1983, a representative of the SED in the Suhl District Council wrote a report to Minister Reichelt describing an incident in which he proposed the "socialist solution" to pollution – planting pollution resistant species in the Thuringian forest – to a group of residents. It did not go over well, he wrote: "Suddenly a mother of three children stood up, angry, and asked the members of the council, if they wanted to offer the children some smoke-resistant lungs."[56] Clearly, SED officials were tone-deaf to the real concerns that East Germans had about pollution – they were not worried as much about trees sickening and dying, as about sickening and dying themselves.

Despite official censorship, doubt about the regime's standard line regarding pollution began to emerge in literature, such as Hanns Cibulka's novel *Swantow*, which expressed concerns about the risks of nuclear energy, Lia Pirskawetz's *Der stille Grund* [Quiet Ground] which dwelt on landscape conservation, and Christa Wolf's *Störfall* [Accident], which had to do with Chernobyl.[57] Some documentary films were able to avoid censorship and be released to audiences in the GDR as well, including Günther Lippmann's 1981 *Goldgruben* [Gold Mines], which was made with the

55 Dieter Halbig, "Weitere Maßnahmen für bessere Leipziger Luft?" *Leipziger Volkszeitung*, November 12, 1987.

56 BArch DK 5 188, "Letter from Klaus Höpcke, minister for culture to Hans Reichelt," August 19, 1983.

57 Axel Goodbody, "German Ecocriticism: An Overview," in *The Oxford Handbook of Ecocriticisms*, ed. Greg Garrard (Oxford: Oxford University Press, 2014), 547–559.

support of the Karl Marx City Council which was "very progressive" for Lippmann. This 22-minute documentary focused on waste and its ecological consequences, and despite some censorship, it was still seen by millions of East Germans, because of the subject – and also because it was screened in theaters as a short film just before the feature film *Lady Chatterley* in the movie theaters. In 1986, Lippmann made a similarly short documentary called *Kostbares Nass* [Precious Water] which focused on water pollution in the Elbe and Werra rivers. The SED allowed this film to be shown because they felt it purported to document advances in the GDR's anti-pollution efforts, even screening it for the West German SPD delegation that came on an official visit in 1987.

Once Lippmann wanted to make a film about the environmental damage caused by air pollution in the Erzgebirge, he ran into roadblocks – it took him four years to receive authorization to film, and the film he made, which involved interviews with forestry officials and citizens who were engaged in the burgeoning ecological movement within the church, was ordered to be changed eight times, and in the end, was never allowed to be screened in the GDR. Censors objected to the fact that it "did not reflect the environmental policies of the GDR" and that it "did not emanate optimism," it was "wrong on environmental policy," and the "members of the State Forestry District did not appear in a leadership role as socialist cadres."[58] It was shown for the first time on TV in November, 1989, after the Fall of the Wall.

Increasingly shut out of the public sphere controlled by the authorities, ecological citizens' groups created new public spaces, often under the aegis of the Protestant Church, where they could discuss environmental questions openly, without being shut down by the Stasi. They published articles in their own reviews, and disseminated these in their own meetings. They used words like "*Waldsterben*" which had been officially banned as "alarmist." One underground journal was called the *Briefe zur Orientierung im Konflikt Mensch-Erde* [Letters Towards an Orientation in the Conflict Between Man and Planet Earth]. It was in the February 1983 issue of *Letters*,

58 DRA 118, DEFA Studio für Dokumentarfilme, "Kostbares Naß," Film No. 552. See also DRA 1-Z 1481.

that the topic of *Waldsterben* in the Erzgebirge could finally be openly broached. In an article entitled "Is the Forest in the Erzgebirge Dead?" the question was finally raised: "Why is it that there is a big environmental problem in the GDR and yet so little public information and debate on the subject?"[59] Subsequent issues continued to explore the subject, with titles like "Origin of the forest decline."[60]

Perhaps no event was more indicative of the strength and creativity of the developing grassroots environmental movement in the GDR than the "One mark for Espenhain" action. The Espenhain power plant, near Leipzig, was a notorious polluter in the region. In July 1987, during a Christian environmental seminar in Mölbis, a town near Espenhain, the organizers invited the plant's director and environmental protection officer, as well as the leader of the Environmental and Water Management Council for the Leipzig District. None of the officials present admitted any responsibility for the pollution caused by the Espenhain plant. They did not deny that there was damage to surrounding vegetation, but they claimed that they were unable to do anything meaningful about it because of decisions which came from a higher level.[61]

Furthermore, during the discussion, the participants learned that the power plant no longer had anti-dust filters. The official reason had been that such filters were too expensive, but reality what was going on was that already as early as 1965 the authorities slated Espenhain – a very outdated facility – for closure, and beginning in 1975 it therefore had no more access to investment. And yet, because of the energy crisis, it continued operating out of necessity. Rather than simply passively accept that as an answer, the citizens groups began to formulate a solution, or at least, an active response. In 1988, the Christian environmental group Rötha, with help from other grassroots environmental groups in Dresden, initiated the action "One Mark for Espenhain" (*Eine Mark für Espenhain*). "One Mark for Espenhain" sought to gather donations, one mark at a time, purportedly

59 Ursula Salzmann and Joachim Krause, "Stirbt der Wald im Erzgebirge?" *Briefe* 7 (February 1983).

60 Bernd Kaina, "Ursachen des Waldsterbens," *Briefe* 8 (November 1983).

61 Roland Quester, "5. Umweltgottesdienst," *Streiflichter*, July 1987.

to buy a filter for Espenhain. In reality, there was little chance of the action actually collecting enough donations to buy the filter (especially at the rate of only one mark per donor) – but that was not actually the point. The point was to collect signatures, which the campaign did each time someone donated. Collecting signatures for petitions was not legal in the GDR, but donation drives were, and so the group collected signatures as records of donation receipts (*Quittungen*) – a testament to the creativity and wryness of the ecological grassroots groups, as well as their ability to use ecological protest as a platform for a wider critique of a lack of freedom of speech. Ultimately, 100,000 citizens donated – and therefore signed – in the action. The Espenhain action was the most well-known of many actions taken by grassroots groups, and it followed a typical pattern, in which the authorities would try, and fail, to convince the population, and then environmental and Christian groups would respond by taking a leadership role in actually looking for a solution and informing the public.[62] They, in effect, created their own, new, public sphere.

The environmental grass roots groups knew that their information had a better chance of being disseminated in the GDR if it first was smuggled into the FRG, and then broadcast from there, where East Germans could receive radio and TV signals (or published, and then smuggled in to the GDR). Several of these groups made surreptitious videotapes of environmental pollution in the GDR, and had them smuggled into the FRG, to be broadcast on West German news, for example on the television magazine *Kontraste* on Channel 1 (ARD) in West Germany.[63] One of these videotapes was about the dangers of uranium mining in the GDR; another one was the factory town of Bitterfeld, and this one had an impact that resonated widely in the FRG and in the GDR. It was filmed by the East Berlin ecological group "Arche Nova" with the help of a West Berlin journalist, who smuggled the videotape out of the GDR (West German journalists, like West German diplomats and members of parliament, were not searched at the German-German border). The video footage was broadcast on September 27, 1988 as a documentary called *Bitteres aus*

62 Interview with Karl Weise, Pegau, July 24, 2000.

63 BStU ZAIG, Vi, Nr. 590, ARD-Magazin "Kontraste" VHS.

Bitterfeld [Bitter Tidings from Bitterfeld], and was seen in the GDR as well, and had the intended impact among East Germans. One *Eingabe* from a resident of the Saxon town of Freiberg, sent to the Ministry of Environmental Protection, on October 2, 1988, read: "It is known that in the Freiberg area a number of measurements were performed. Why were the results and conclusions not communicated to the citizens?" It continued: "Furthermore we find it very regrettable that information [about environmental pollution] only reaches people in the GDR through western media and never through their own press. I conclude from this that you are either powerless in such matters, which I suspect, or else you must have no knowledge of these things."[64] The documentary caused enough concern in the SED Central Committee that they began looking into the matter. On October 5, 1988, a report from the Raw Material Industry Section of the Central Committee was presented to the Central Committee. Prepared with the help of local District Councils as well as the Bitterfeld Chemical Factory, it did not take responsibility for the pollution; instead gave counter-arguments about data and insisted that the situation in other countries was worse.[65]

In May 1989, the grassroots ecology group Arche-Nova conducted an action in which they distributed 20,000 letters throughout East Berlin. The recipients were chosen arbitrarily from a GDR telephone book. Once again, environmental activists in the East made use of the proximity of the border and the strong environmental movement in the West: the letters had been printed in West Berlin, and smuggled across the border into the GDR on the person of a Green Party member of Parliament, Karitas Hensel, who had diplomatic immunity as a member of the Parliament. In the evening of May 9, 1989, the letters were delivered. These letters contained information about forest damage, air pollution in the GDR and a complaint

64 "Ohnmacht oder Unkenntnis?" accessed August 8, 2017, <https://landesarchiv.sachsen-anhalt.de/fileadmin/Bibliothek/Politik_und_Verwaltung/MI/LHA/externa_alt/89_06/8990_Juni_Protest_9.htm>.

65 "Bitteres aus Bitterfeld," accessed August 8, 2017, <https://landesarchiv.sachsen-anhalt.de/fileadmin/Bibliothek/Politik_und_Verwaltung/MI/LHA/externa_alt/89_06/8990_Juni_Umwelt_3.htm>.

form to be sent to the government.[66] Even though the action was partially hindered by the Stasi, who had known about the plot since March thanks to an informant (*Inoffizieller Mitarbeiter* [Informal Cooperator] or "IM"), enough of the letters were delivered that it sparked a reaction from the authorities. An interview with the philosopher Alfred Kosing, professor for dialectic and historic materialism in the Institute for social sciences at the Central Committee of the SED, was published in the West German left-wing newspaper *Unsere Zeit* (which had long been subsidized by the SED) on May 27, 1989, in response to the letters and entitled "A Critique of Greenpeace's Ignorance." For Kosing, Greenpeace ignored the environmental policies of the GDR against air pollution. No technology was available to counter sulfur emission, he claimed, and there remained still much research to be done on that score. When confronted by the interviewer with data showing the true level of air pollution in East Germany, specifically using the example of Dresden in which pollution had reached 3000 ug/m^3, Kosing admitted openly that "the sulfur dioxide pollution in GDR ... is already extremely high. It might well be the highest in the world." On the issue of environmental data being made a state secret, Kosing answered that he had "never seen such an ordinance." When asked specifically about residents having access to data about pollution levels in their area, which would be of immediate importance to them and their health, Kosing insisted that "the public data is accessible" but then diverted the question by stating "*what is decisive is not to publishing every piece of data, but rather solving the problem*."[67] The action had been a success simply by drawing the East German press into a debate in the public sphere about environmental damage, which had until then been a state secret and only for discussion behind closed doors in Party circles.

Once they realized they had been dragged into the public sphere, the authorities realized they could no long simply ignore the problem of pollution, and set to work organizing a series of concrete responses. A report from the Stasi from 1987 claimed that a press offensive was needed

66 See Uwe Bastian, *Greenpeace in der DDR* (Berlin: Edition Ost, 1996).

67 Bernhard Müller, "Kritik an Unwissenheit von Greenpeace," *Unsere Zeit*, May 27, 1989.

because the citizens were beginning to lose trust in their government. But very quickly the various organs of the state realized how difficult a coordinated and effective public response would in fact be. A report from the Dresden District Council to Minister Reichelt in November 1989, shortly before the Fall of the Berlin Wall, highlighted the difficulties of engaging in a public relations campaign to restore trust among the people regarding pollution, claiming that any such work done by local authorities would be difficult, because no actual results had been seen – the population had been living in pollution for so long that mere words were not going to change their mind, not even promises from the Council of Ministers in Berlin.[68]

The Academy of Sciences authored a report on April 18, 1989, which underscored the dilemma that the GDR had put itself in regarding pollution. The citizens were worried and confused by "damage" and "ecological danger," it stated, and emphasized that an environmental consciousness was starting to emerge because of the discussions in the public sphere. The state needed to take control of this discussion, it argued. Professor Dr. Wolfgang Mundt, Director of the Scientific Council for Fundamental Environmental Engineering and Environmental Protection to the bureau of the Academy of Sciences, proposed the creation of a Scientific Information Center for the Environment alongside the Environmental Council.[69] The goal of the proposal ultimately was to provide scientific information to citizens, from scientists (who might seem more trustworthy than Party officials or local politicians or factory directors) "by setting up a public 'scientific environmental library,'" rather than leave this role to the ecological movement, which was gaining the trust of the people by filling the void in reliable scientific information that the regime had left wide open. Mundt had in mind especially the environmental library in East Berlin (in the Zions Church in Prenzlauerberg in September 1986 and closed by the Stasi in November

68 BArch DK5 3386, "Konzeption zur schnellen und spürbaren Verbesserung der Umweltbedingungen in den Ballungsgebieten des Bezirkes Dresden."

69 The goal of this scientific council was to coordinate the research on environment inside the academy of science in relationship with the universities and other organization. See Horst Paucke, *Chancen für Umweltpolitik und Umweltforschung: Zur Situation in der ehemaligen DDR* (Marburg: BdWi-Verlag, 1994).

1987) and the Ecclesiastical Research Center in Wittenberg, both of which had emerged as the main sources of data and research on the environment in the GDR – he wanted official scientific bodies like the Environmental Council and the Academy of Sciences to reclaim that role, a role that he believed rightfully belonged to the official scientific community.[70] The goal was to regain control on environmental issue in the public arena.

By 1987, the pressure from both inside the GDR and from outside began to force cracks in the GDR's ban on public discussion of environmental data. The Timber and Forestry Commission of UNECE began demanding real data on forest damage in the GDR, and threatened to publish unofficial *West German* data about East German forests – and that got the East Germans to act, prompting them to send the real data, on June 30, 1987, to the UNECE Timber Commission. Knowing that it would be published by the UN, the Council of Ministers voted to finally make the data public in the GDR, on November 5, 1987. In fact, other Eastern European states had already sent the real data to the UN, including Czechoslovakia and Poland, to be made public to an international audience, and the East German leadership wanted to end its isolation. The data was, comparatively speaking, not as damning as some had thought: it reveal that 37 percent of the GDR's forest, or 1 million hectares, was damaged by pollution. This was actually far less than the situation in the FRG, where 53 percent of the forests had sustained some damage due to air pollution. And yet still, the GDR was lax in making the data publicly known: despite having voted to release it in November 1987, it was not published in the East German print press until May 1988,[71] and crucially, it was not until July 1989 that it was broadcast on television in a documentary about the Erzgebirge and the Thüringer Wald;[72] presumably, authorities were delay-

70 Archiv der Berlin-Brandenburgischen Akademie der Wissenschaften (ABBAW) VA 14052/11, "Empfehlungen des Umweltrates zur umweltstrategischen Standpunkt-Bestimmung aus der Sicht der Grundlagenforschung der DDR, e W. Mundt."

71 Allmuth Schaarschmidt, "Prophylaxe und Therapie für unsere grüne Lunge," *Neue Zeit*, May 28, 1988.

72 DRA AC 2845, "Leben für den Wald: Ein Film des Redaktions Kreisläufe," July 3, 1989.

ing broadcasting what they still saw as highly sensitive information for as long as they could. When they did broadcast, it was the first time since 1982 the GDR showed its weakness in reacting to environmental damage, but not its powerlessness because the articles were always accompanied with a socialist solution.

It was not until October 23, 1989 – five days after Honecker was forced to resign due to the rising protest movement in the GDR – that the GDR press made specific reference to the amount of pollution that had been emitted in the GDR. A dispatch from the *Allgemeine Deutscher Nachrichtendienst* (the central state-controlled news service in the GDR) claimed that 4,990,000 tons of SO_2 and 955,000 tons of nitrogen oxide had been emitted in the GDR. It was the data which was sent to the UNECE for the year 1987, but in fact the real data for the SO_2 levels was 5,560,000 tons and for the nitrogen oxide 820,000 tons.[73] The figures came from interviews with two scientists from the meteorological service.[74] Yet still, until the very end, the GDR tried to keep control over its data in some form. It was only on November 13, 1989, four days after the Fall of the Berlin Wall, that the GDR lifted the designation of "state secret" on environmental data, declassifying and making accessible all its information.[75]

On November 21, 1989, East German TV broadcast a new ecological show entitled *Ozone* which gave a platform to ecological activists like Raimar Gilsenbach and Sigrid Rothe, as well as scientists who could speak about the data on pollution. Yet, even once the data was no longer a secret, most of the news programs that reported on pollution followed the old patterns of shilling for the regime rather than conducting investigative journalism: reporters would interview local authorities, who gave bland reassurances to local citizens. Only in March 1990, did

73 BArch DK 5 2115, "Übergabe von Daten und Informationen über Luftverunreinigungen der DDR an die Wirtschaftskommission der Vereinten Nationen für Europa (ECE) und den Rat für Gegenseitige Wirtschaftshilfe," June 28, 1988.

74 Dr Rolf Bartomek, "Schaden unsere Preise der Umwelt?" *Nationalzeitung*, October 24, 1989.

75 *Gesetzblatt DDR Teil I*, no. 22, 241–242.

this pattern of controlling information really begin to change. Images of pollution, often visually striking, were published, and the headlines came to more closely resemble the truth of the situation. For example, on March 10–11, 1990, the *Ostsee Zeitung* ran an article on village of Mölbis, directly adjacent to (and downwind of) the Espenhain plant. The articled was entitled "People Live Here" and emphasized the fact that the village was constantly shrouded in smoke – and the photographs it put on the cover demonstrated this fact in shocking detail (see Figure 2).

Hier LEBEN Menschen

Fotoreporterin Angret Huth fotografierte in Mölbis bei Espenhain

Figure 2: "People *Live* Here" from the *Ostsee Zeitung*, Spring 1990.

Other such articles followed, each one a revelation for East German residents. Later that month, on March 30, the *Thüringer Allgmeine* (which had been *Das Volk* until January 13, 1990) ran the headline "Bad air and formaldehyde sicken children."[76] It came from an interview with Kurt Menzel, the director of the children's hospital in Erfurt, and took place during a smog emergency in the town. Yet in these articles, journalists and residents were able to grasp the complexity of the situation: the economy and the ecology could not be disentangled from each other. For example, an article in the *Sächsische Zeitung* on April 6, 1990, pointed out that the same factories that created pollution also provided work for residents, and cleaning up the environment likely would mean closing many of the factories, including the synthetic textile factories that had been a mainstay of Saxon industry in the GDR (and before). The article, which ran the headline "Is There a Future for Synthetic Silk? Shut Down for Environmental Protection – 400 Workers Out on the Street?" highlighted the dilemma, especially in a potentially capitalist environment: reducing environmental damage to the forests would mean an increase in unemployment.[77] The difference now was, for the first time, East Germans could openly debate the costs and benefits of favoring industry versus ecology.

Conclusion

Initially, when the GDR was created, there was a place for an open debate on air pollution, at least in the framework of building a socialist society. The discourse involved a belief in technological progress and the patience required to achieve this progress by citizens of the GDR.

76 "Dicke Luft und Formaldehyd machen Kinder krank," *Thüringer Allgemeine*, March 30, 1990.

77 "Hat die 'Kunstseide' noch eine Zukunft? Schluss mit der Umweltverschmutzung – Aus für Arbeitskräfte?," *Sächsische Zeitung*, April 6, 1990.

Later, at the beginning of the 1970s, when the GDR was more committed towards the creation of a socialist consumer society, when the socialist dynamic slowed and where the environment became a political issue, it was becoming increasingly difficult for the authorities to convince its citizens that its policy decisions were correct. In part, this was because the state decided to try to mask the reality of the situation, rather than try to convince the public that the situation was being dealt with in good faith. This left a void for the West German media, aided by the ecological movement in the GDR, to fill with their own narrative, a narrative which generated doubts about the capacity of the regime to fight the consequences of air pollution. It was in this way that the policies of the GDR state and Party were first delegitimized and then repudiated in November 1989.

GERNOT WALDNER

Ecology and its Discontents: The Concept of Nature in Elfriede Jelinek's *Oh Wildnis, oh Schutz vor ihr*

Introduction

Austrian politics in the 1980s was characterized by democratic improvement and ideological polarization. After almost 20 years of social-democratic rule, the electorate became increasingly dissatisfied with the center-left government, a discontent which was voiced in two events. In 1984, a state electricity supplier planned to build a power plant in the ecologically fragile wetlands of Hainburg, situated outside of Vienna, leading to the occupation of the Hainburger Au. This occupation was the first case of civil disobedience in the Second Republic. The momentum behind the squatters was caused by a growing ecological movement and fostered by criticism of the nepotism within the ruling left-wing elite that commissioned the power plant. Subsequently, this gathering prompted political referenda, as well as the foundation of the Green Party, whose members would first take their seats in parliament two years after their formative sit-in in the wetlands.[1]

In 1985, one year after the successful preservation of the wetlands, Elfriede Jelinek published *Oh Wildnis, oh Schutz vor ihr*, a text that returns to the experimental style of writing she had developed at the beginning of her career.[2] Thereby Jelinek partakes in an experimental

1 Franz Schandl and Gerhard Schattauer, *Die Grünen in Österreich: Entwicklung und Konsolidierung einer politischen Kraft* (Vienna: Promedia, 1995), 23–78.

2 Elfriede Jelinek, *Oh Wildnis, oh Schutz vor ihr* (Reinbek bei Hamburg: Rowohlt, 1993).

transformation of the genre "Heimatroman," joining writers like Gert Jonke and Franz Innerhofer.[3] It is difficult to discern how exactly Jelinek perceived this occupation but in an interview conducted by the *Kölner Stadt-Anzeiger* she states that she tried to find an aesthetic form for recent political issues. Elaborating on these issues she mentions the failure of Social-democracy as well as the primarily emotional connection between environmentalists.[4] Indeed, the green movement lacked political analysis and coherent ideology, part of the movement was even considered right-wing. Jelinek omits the democratic achievement of the referendum and instead criticizes the new political coalitions enabled by ecological movements, probably because she did not see anything positive in the new green coalition which presented itself to the public as viable political movement. In the media, it was one event which illustrates Jelinek's estimation of the wetland squatters. On a somewhat jubilant occasion, the occupiers of Hainburger Au presented themselves to the media in an event which became known as "Die Pressekonferenz der Tiere" [The Animals' Press Conference] (see Figure 3). Left-wing writer Peter Turrini, member of the Communist Party, entered the stage dressed as a toad together with the chairman of the young conservatives, Othmar Karas, a cormorant on this day.

The individuals on stage formed a coalition unthought of, ideologically and biologically. Additionally, to the conservative and the communist representative, the movement against the power plant was substantiated by a petition in order to provide the legal framework to prohibit the erection of the power plant. The petition's initiator was Konrad Lorenz, a man of scientific fame and political infamy, winner of the Nobel Prize in 1973 and officer for racial policy during the Nazi regime.[5] Lorenz completes this ecological assemblage of many kinds.

3 Andrea Kunne, *Heimat im Roman: Last oder Lust? Transformationen eines Genres in der österreichischen Nachkriegsliteratur* (Amsterdam: Rodopi, 1991), 3–16.

4 "Schreiben in der Männerwelt, Suche nach Formen für politische Inhalte," *Kölner Stadt-Anzeiger*, Nr. 280/14.

5 Ute Deichmann, *Biologists under Hitler* (Cambridge, MA: Harvard University Press, 1996), 179–206.

Figure 3: The "Animals' Press Conference" at the occupation of the Hainburger Au.[6]

The unity of this movement can hardly be described in terms of a coherent ideology, but rather, as Jelinek puts it, in a broad emotional opposition to the government or in a sentimental feeling for wetlands. Indeed, the rising ecological movement was far form having a party line at this juncture. What was clear was that this new movement would attract people from the left like the aforementioned writer Peter Turrini, and thereby represent a political competitor for established parties. In consequence, the Left might face difficulties to counter a resurgence of the Right.

An opposition to the Right became pertinent around this time, for the second outstanding event in Austrian politics of the 1980s was the election of former SA officer, Kurt Waldheim, as president of the Second Republic. When the electorate voted for Waldheim, the Austrian role in the Third Reich, which had largely been neglected since 1945, eventually

6 <http://www.protestwanderweg.at/au/au_02.php> (accessed July 13, 2017).

became more of a public concern. Years before his election, Jelinek and other intellectuals had already thematized the myth of Austrian victimhood during World War II, yet Waldheim's presidency epitomized Austria's distorted view of its past, inspiring Jelinek to write political texts such as "In den Waldheimen und auf den Haidern."[7] The contrast between initial and ultimate reactions to Waldheim's election demonstrate the impact on the national self-image. When journalists and the Jewish World Congress instigated the investigation into Waldheim's past, prominent figures of the political center at first opposed scrutiny. Bruno Kreisky, former prime minister, acclaimed initiator of social reforms, and himself of Jewish origin, defended Waldheim against the "Jews abroad," prolonging the sentiment of provincial patriotism against voices from outside. Contrary to Kreisky, his successor, Kurt Sinowatz, revoked the official position of the Social Democratic Party after the evidence against Waldheim proved to be conclusive. Sinowatz's ironic sentence on the matter became part of collective memory: "Ich stelle fest, daß Kurt Waldheim nie bei der SA war, sondern nur sein Pferd."[8] For the context of this novel, it is important to note that Waldheim's election was at first defended by the center-left establishment, for major parts of the Social Democratic Party would have also had to confront their neglected past. To conclude this sketch on the political landscape of Austria in the 1980s, the democratic improvement as instigated by the occupation of the Hainburger led to a form of stasis within the Left since many of their members joined the ecological movement in spite of its political blend reaching from communism to former Nazis. Jelinek perceived this form of stasis as threat to the overall situation as an organized resistance against the resurgent Right would become difficult if the Left would continue to lose support. Therefore, the ecological movement led to a political wilderness.

7 The title makes a pun exploiting the ambivalence between the Austria's internationally best-known politicians and rural landscapes, heath and forest. The full text is available online: Elfriede Jelinek, "In den Waldheimen und auf den Haidern," last modified July 15, 2017, <http://www.elfriedejelinek.com/fboell-p.htm>.

8 *ORF Mittagsjournal*, March 11, 1986.

However, Elfriede Jelinek does not take sides in political debates. Although she had been a member of the Communist Party since 1974, she personally confirmed what literary critics in spite of the controversial reception of her work would have agreed on, that her books could not easily be utilized as propaganda for the Communist Party.[9] Nevertheless, politics was always one of Jelinek's major concerns and increasingly so in the 1980s as she established herself as a public persona. She began to speak at public events and published articles on political issues in left-liberal periodicals such as *Profil*, *Falter*, and *Der Standard*.[10] This engagement, together with the numerous references to contemporary events in her works, renders it difficult to understand her writings in separation from politics. Asked about her agenda as a writer, Jelinek concurred with the Marxist demand to change consciousness of people in order to engender societal change. So, the question arises: how could her work serve political ends while not being propaganda but widely considered vanguard literature? In the same interview, in which Jelinek rejected her works be a medium for political messages, she elaborated on her poetics by saying that she aims to find an adequate form for the political content she addresses in her texts.

The overall style of *Oh Wildnis, oh Schutz vor ihr* is brash, curt, and impudent, refusing to provide the reader with a coherent flow of reading. For a text programmatically dealing with nature, this is already a statement in itself, as one would expect rather flowing descriptions covering the wide range of flora and fauna instead of short mechanic sentences unwilling to connect to each other. The form of the text emulates a wilderness, in spite of the content being concerned with nature. The entire text is written in this impertinent *genus humile*, assembling proverbial malapropisms, rural idioms, and pithy sarcasm. These choppy bits and pieces of language are the means as well as the target of Jelinek's political intervention, because, according to her, their form is a statement on the political aspects their content deals with.

9 "Interview," *Bio-Technik*, May 1983.

10 For a general overview on Jelinek's life around this time: Marlies Janz, *Elfriede Jelinek* (Stuttgart: J. B. Metzler, 1995), 99–111.

The nature of Jelinek's critique of "nature" is therefore not systematic in a Marxist sense: ecological policies, as they become known through the occupation of the Hainburger Au, are not exposed for their inherent contradictions. As mentioned, the text evades features that could be described as linear or coherent. Accordingly, the first reviews of this book did not fail to note that this text does not have a plot, that it does not come with a consistent narrative perspective, that it is neither dystopian nor utopian, and that it does not suggest that anything positive can be said about contemporary society.[11] Even the publishing house Rowohlt seemed to struggle in finding an apposite label for this text: the cover of the book says "prose" ("Prosa"), clarifying that one can expect neither rhymes nor dialogues on the pages to follow, which is both untrue.

Apart from the few characteristics of this book we already mentioned, scholars struggled to find positive characteristics for this text. If the political intervention of this text lies in its form, then it was not clear what this form aimed to achieve. Years later, Jelinek referred to a biological process to state the goal of her critique: she would like to make her readers vomit. In the following, I would like to take up her emetic cue and scrutinize formal aspects in *Oh Wildnis, oh Schutz vor ihr* that could cause an involuntary expulsion of content. My overall argument is that Jelinek critiques the ramifications of the ecological movement because it does not account for manners of speaking about nature as they already exist in ordinary language. For Jelinek as a writer, the ecological

11 There are few scholarly articles on this book: Bärbel Lücke, *Semiotik und Dissemination: Von A. J. Greimas zu Jacques Derrida: Eine erzähltheoretische Analyse anhand von Elfriede Jelineks "Prosa" "Oh Wildnis, oh Schutz vor ihr,"* Epistemata. Reihe Literaturwissenschaft; Bd. 409 (Würzburg: Königshausen & Neumann, 2002), 154–203; Gerald Chapple, *Towards the Millennium: Interpreting the Austrian Novel, 1971–1996: Zur Interpretation des österreichischen Romans, 1971–1996*, Studien zur deutschsprachigen Gegenwartsliteratur; Bd. 10 (Tübingen: Stauffenburg, 2000); Heike Weinbach, "Volkstöne und Plastikfladen: Elfriede Jelineks Satire 'Oh Wildnis, oh Schutz vor ihr,'" *Augen-Blick: Marburger Hefte zur Medienwissenschaft* 5 (1988): 78; Rita Svandrlik, "Zur Typologie der Konflikte in Elfriede Jelineks *Die Liebhaberinnen* und *Oh Wildnis, oh Schutz vor ihr*," in *Akten des XI. Internationalen Germanistenkongresses Paris 2005* (Bern: Peter Lang, 2007), 77–83.

movement cannot replace the entire discourse on nature but only add authority to preexisting occurrences in language which are derived from nature too. This does not imply that ecological assumptions of that time are adequately represented in ordinary language but that Jelinek's fear was that the valorization of ecology might reinforce existing references to nature which she politically rejects. In other words, Jelinek assumes that even if ecological politics were to become successful, it would be unlikely that most people would gain an acumen similar to environmentalists and that their usage of sayings relating to nature would be limited to environmentalism. Her novel rather suggests that with growing success of the ecological movement, nature might become a value in itself, affecting more than squatted wetlands. Jelinek's manifold usage of "nature" therefore demonstrates how we also refer to natural phenomena or use metaphors derived from nature when we speak about tourist sites, gender politics, TV documentaries, romantic poetry, conservative values, and, in some cases, ecology. This has to be kept in mind when I talk about "nature" in the following. Consequently, the text opposes a monolithic view of nature, the one employed in the press conference of animals, and instead battles the wilderness of existing usages by means of literary form.

I am going to discuss three rhetorical strategies Jelinek employs throughout this text in order to challenge a one-sided view of nature and to add an emetic flavor to existing manners of speech. First, I show how Jelinek puts the difference between nature and culture at stake, calling firm distinctions into question in order to render nature a matter of politics again. Second, I discuss how nature distorts a conceptual understanding of countryside tourism as well as social injustice. And third, I expose discrepancies between a cyclical time model prevalent in discourse on nature and a somewhat linear model that seems necessary for political action as well as commemoration. What all strategies have in common is that they force readers to reconsider metaphors they have grown accustomed to, to undo linguistic habits which govern their consciousness. In this way, the form of this text aims to prompt a re-evaluation of the political ramifications instigated by Austria's ecological movement of the early 1980s.

Retrieving Nature

Beginnings of literary texts are often scrutinized because they exemplify the form of a text. In the first couple of sentences, the parameters are set out on which the narrative of the entire book builds upon. Usually, the *caesura* beginnings create between the worlds inside and outside the text consists of the introduction of time, place, and protagonists.[12] As for Jelinek's book on wilderness, the beginning is of particular interest because many scholars characterized this text as one lacking characteristics. Indeed, *Oh Wildnis, oh Schutz vor ihr* does not disappoint its readers, shunning all positive features one would expect in the opening paragraph:

> An keinem besonderen Tag vor dem Wintergefrier geschieht es endlich, es zieht einer sich festlich an (wie er sich das so vorstellt!), er singt nicht was ein Wanderer singt, während er kräftig atmet. Er steigt in den Hohlweg ein.[13]
>
> [Not on a special day before the first frost, it eventually happened, one dresses up in a festive way (as he imagines it, at least!), heavily breathing, he does not sing what a wanderer used to sing. He enters the hollow way.]

Granted that Jelinek's text conforms with literary beginnings insofar as we learn something about time and place, the particular form renders this beginning nevertheless an exception. On a day before the first frost not worth mentioning, something eventually happens. In the first main clause, we don't learn what actually happens but it happens eventually, as if the beginning would be a superfluous event everyone had long expected to occur. Similarly vague, the protagonist appears as an indefinite article, a male someone whose conception of festive dress seems to be misguided, as the insertion in the tone of neighborly gossip suggests. Grammatically, the position of the protagonist is weakened by his role as appositional nominative, whereas a so-called

12 Mikhail Mikhaĭlovich Bakhtin, *"Forms of time and of the chronotope in the novel"* in Bakhtin, *The Dialogic Imagination: Four Essays* (Austin: University of Texas Press, 1981).

13 Jelinek, *Wildnis*, 7.

pseudo-subject, "es," serves as undefined agent of the sentence, diminishing the protagonist and depriving him of agency. We then read the protagonist does not have enough air to sing, which negatively alludes to romantic prose of the German tradition, most notably Joseph von Eichendorff's *Aus dem Leben eines Taugenichts*, in which a virile man singingly sallies into nature.[14] In contrast to the romantic topos, this stranger seems to be physically strained, even if we yet don't know why he is breathing heavily, living in a natural surrounding appears an arduous task. After the first two sentences, we can only infer from the dialectal word "Wintergefrier" that the diegetic world might be located somewhere in the Southern-German or Austrian countryside. And there, in this unknown region, notwithstanding, evoking lakes and forests, somewhere there, the usual happens. This beginning is deliberately insignificant, place and time could be replaced with any other year and the male protagonist is an everyman from a German speaking countryside.

At the end of this chapter, more than 80 pages later, the same everyman still stands in the *Hohlweg*, singling out the importance of this sunken path or hollow way. The sunken path serves as crucial concept in this text in order to thematize the relation between nature and culture. Sunken paths are culturally used by men like our protagonist and naturally indurated by rain and snow. As the protagonist never leaves the sunken path, the text continually points to the entanglement between nature and culture yet the roles are syntactically reversed, as the usage of the pronouns and the pseudo-subject has shown. In this sunken path, the agency stays with the natural part instead of the human activity. The creaturely part of our protagonists causes the heavy breathing and what happens to him and his body seems to be a part of the yearly cycle as much as summer and winter are part of larger natural order. On a literary level, the allusion to romantic writers does not create a semantic surplus but evidences the discrepancy between a romantic

14 On the usage of negations in Jelinek's prose: Rita Svandrlik, "Gestörte und umgekippte Idyllen: Zum Komplex 'See' in Adalbert Stifters *Hochwald* und Elfriede Jelineks *Gier*," in *Idyllik im Kontext von Antike und Moderne: Tradition und Transformation eines europäischen Topos*, eds. Nina Birkner and York-Gothart Mix (Berlin: de Gruyter, 2015), 300. For this particular reference see: Joseph von Eichendorff, *Aus dem Leben eines Taugenichts* (Frankfurt/Main: Suhrkamp, 2007), 5–6.

view of nature and the physical strain of human beings living in the countryside. Employing literary allusions, Jelinek's opening creates a gap between our romanticized sentiments and the strenuous life of the everyman. Even more so, she uses the allusive surplus to negate the possibility of an individual story worth telling and neither would the unknown man be able to express himself properly, as the social oppression in brackets indicates. Erich, whose proper name we only learn dozens of pages later, does not have a story to tell or one worth being told.

The beginning of this text thus suggests that nature is not as recent an entity as forged by Austrian politics in the 1980s. Understood as rural environment, nature shapes the lives of its inhabitants since ages as reason for their toil. Rather than an area worth being protected, nature is the active force determining the arduous activities of lumberjacks like Erich. Jelinek shifts the understanding of nature from a passive object worth being protected by ecologists and about to be exploited by a Social Democratic elite back to the exploitation of workers. Her text makes nature an issue of the political Left again. Accordingly, she states in an interview that the Social Democrats are no longer the party caring for workers.[15] Jelinek employs references to romantic prose to do this, demonstrating that our perceptions of nature are formed by cultural patterns and that what concurrent political discussions view as "nature" stands in opposition to the interests of the lower class on the countryside.

Notions of Nature

The opening section of this text corroborated that nature and culture are always intertwined yet the distribution of agencies contrasts with the cultural tradition we became accustomed to. On many other occasions, Jelinek's book further expands on misguided conceptions of nature by probing semantic

15 Elfriede Jelinek, "Zum 27. Parteitag der KPÖ: Der Umweltschutz dürfte nicht den Grünen überlassen werden," *Aus der Parteitätigkeit*, April 1989, 13.

fields in which metaphors derived from flora and fauna are employed. In two of these fields these metaphors serve to represent something as unique and the commercially most important of these two fields is tourism. By the 1920s the leisure industry had coined their own image of nature in order to supply mass tourism with labels that would advertise the countryside to city dwellers.[16] Rhetorically, the unique selling point of a certain place was fashioned from proper nouns, deictic expressions, and extensive descriptions. Phrases like "this idyllic lake" and "that majestic mountain" aimed to create an aura for central European geography. Aesthetically, these descriptions regress into naturalism, a literary style Jelinek explicitly opposes. In addition to the labeling of regions, the development of mass tourism encouraged a kind of patriotism for ones area, creating a mutual amplification between the competition for tourists and the penchant for regional pride. Jelinek's text ridicules this economically fostered aspect of provincial hubris: "Tirol ist fast wie steirisch."[17] The definite noun sounds as specific as the derived adjective. By juxtaposing noun and adjective, the text derides the description of natural sites, for these tourist sites are like many others and their self-esteem results rather from market pressure than from an original quality.

On a semiotic level, Jelinek's rhetoric strategy to reveal the causes behind a mythification of nature draws on Roland Barthes' work on everyday mythologies. Barthes defines the mythos as a story whose function is to clear a historical event from its contingencies in order to naturalize it and preclude critique.[18] Jelinek's text is the counter program to the process Barthes describes in "Mythologies": she re-politicizes issues which are perceived as matter of fact.

However, Jelinek's text does not limit itself to cultural critique, it also changes perspective and points to the natural features of a space which cannot be embellished by words and pictures. Among those natural aspects, it is mostly geographic features which determine the possibility of touristic success for a given area. "Die Gegend ist so ungeschickt, daß man ihr zahlenmäßig

16 Ursula A. J. Becher, *Geschichte des modernen Lebensstils: Essen, Wohnen, Freizeit, Reisen* (Munich: Beck, 1990), 153–225.

17 Jelinek, *Wildnis*, 87.

18 Roland Barthes, *Mythologies* (New York: Hill and Wang, 2013), 254.

nur wenige Gäste schickt. Kein großer See."[19] In the second sentence, the text laconically captures the sine qua non for successful summer tourism. Nobody would pay for hot summers in an area where one could not refresh oneself in cool water. The caustic aspect of Jelinek's sentence stems from the sentence without a verb. A sentence requires a verb to convey an action as much as an area requires a lake to attract tourists. Moreover, the absence of the verb reflects the fact that nothing can be done against this geographic deficit, for Austrian and Southern German lands have so many lakes that an artificial lake would be an unprofitable investment given the abundant existence of ponds in regions nearby. The first sentence, however, anthropomorphizes the region as "ungeschickt," continuing the metaphorical realm according to which a region "has" geographical features and "offers" opportunities for vacation. Of course, the anthropomorphic usage with regard tourism is limited to positive aspects. Jelinek is more consequent in working this semantic field and thereby sheds light on the artificiality of the entire anthropomorphism. She deliberately picks an adjective such as "ungeschickt" which is commonly used to disdain girls, women, or service personnel. The sexist usage equals a disqualification for a house wife whereas the touristic usage implies to discharge someone from his or her position. So, these two short sentences work through different layers of metaphors derived from nature. First, they disqualify a region in an anthropomorphic way, which is the same way in which a region is romanticized for touristic purposes. Second, they extend the anthropomorphic metaphor by adding negative qualities typically used for women in order to outline the geographical features a natural landscape depends on. And third, they single out the artificiality of both usages: the fatalistic character of a judgment typically passed on women and the specious innocence with which an area is described to render it a profitable region.

As we have seen, the text discusses how discrimination affects males as well as females. In the first paragraph, a gossipy voice mocked the lumberjack for not being able to express himself as an individual, in addition to his deteriorating toils. Although Jelinek begins her text with an example of how gossip, stereotypically a female sphere, predominates over a man,

19 Jelinek, *Wildnis*, 7.

the discrimination of women plays a more important role in her text, not least because the normative description of women borrows vocabulary from nature since antiquity.[20] A short paragraph on female flowers demonstrates some aspects of the feminist critique in this book:

> Dieses Gewächs erdient den schönen Namen Enzian nicht. Auch Selbstzeitlose paßt nicht zu ihm. [...] Wie liebte dieses spezielle Kind Natur, der Sohn liebte die Technik mehr.[21]
>
> [This plant does not deserve the name *Enzian*. Also "spring crocus" does not fit. [...] How this peculiar child adored nature, the son preferred technology.]

By omitting the initial fricative, "verdienen" becomes "erdienen," and Jelinek suggests that not all genders do deserve a symbolic choice between different kinds of flowers. Although women have been compared with flowers, no woman could dare to become an *Enzian*, still the epitome of an alpine flower for every hiker, the emblem on Austrian coins, and legally protected by nature conservation. But neither does the neologism "Selbstzeitlose," a combination of the flower "Herbstzeitlose" and a person's lifetime, seem to befit women from the countryside because the suffix "-lose" signifies someone who is in charge of his or her time. Jelinek's feminist critique therefore highlights the lack of self-determination women were facing in uneducated areas. As regards the tradition of floral names for women, the text refrains from mentioning those types of flowers which have been plucked for wooing, complimenting, and condescending. Nevertheless, the rhetoric of this floral tradition is present. Furthermore, Jelinek demonstrates her acumen in the idiom of the countryside by saying "child" instead of "girl," because in many regions, including the one I come from, the name of the species is still in use for female children, diminishing their social status in comparison to males. The second sentence concludes the discrepancy in terms of autonomy between genders: this particular child loved nature whereas the boy became interested in technology. The adjective "spezielle" conveys bitter irony as

20 Publius Ovidius Naso, *Liebeskunst – Ars Amatoria* (Berlin: Akademie Verlag, 2011), 79–80.

21 Jelinek, *Wildnis*, 20.

there is nothing special about the connection between women and nature since some features of social life are not considered particularities but commonalities. In short, Jelinek's use of language asserts that the language of nature perpetuates the inequality between genders.

Jelinek's second formal critique on the discourse around nature mixes understatement and exaggeration, suggesting that natural metaphors are either too abstract or too concrete in order to use them in a politically meaningful way. In the case of landscapes, they are too concrete, for they take an area as unique whose economic value hinges on a number of geographical aspects abstracted from it. In the case of women, they are too abstract for they perpetuate stereotypes which subdue individual freedom. Taken together, the form of Jelinek's prose contests the political significance of metaphors derived from nature because these metaphors are deficient in their conceptual quality addressing societal issues. If these metaphors derived from nature contain their own political agenda, as Jelinek formally shows in her sarcastic wordplays, it cannot be a political goal of the Left to strengthen the importance of nature in general since nature proves to be a wilderness of usages, ranging from sexist paternalism to economic mystification.

Breaking the Circle of Life

The last rhetorical strategy I would like to discuss concerns the implications of a holistic eco-system in which every part, be it animal or plant, depends on every other part. Given the intricate entanglement of nature and culture, as outlined above, the continuing dissemination of the idea of an ecological system threatens to impede political action because all cultural parts connected to the realm of nature would be considered worth preserving. Jelinek puts it bluntly: "Alles darf so bleiben wie schon einmal etwas genauso gewesen ist."[22] In this paraphrase of the circle of life, the polemic tone stems from the

22 Jelinek, *Wildnis*, 58.

particles. "Schon einmal" and "genauso" expose the conservatism inherent to a concept of nature targeted at preservation while precluding change. However, as I tried to show, nature has many meanings in this text. It does not only refer to parts of society in which preservation would cement the status quo but also to an expanding tourism for which the countryside is modernized and frequented by visitors. Contrary to chauvinism, tourism does not preclude change but transforms rural societies socially as well as culturally. But what does being part of a larger incoherent whole imply for the concept of nature?

Jelinek seems to be well aware of the transformative force of capitalism as it unfolds on camping sites and in hotel lobbies. In connection with the policy of preservation, nature becomes paradoxical when seen as economic endeavor. "Die Natur erhält in diesen Jahren ihre Endform."[23] Nature always seems to have its final form yet this final form is a projected one because just before this sentence the text thematizes road construction and other infrastructure expansion which makes the region easier accessible for tourists. Apart from other usages, these two notions of nature seem to contradict each other. In the first sentence, nature is an ecological system which deserves to be protected. In the second sentence, nature refers to a business for which rural areas need to be transformed. How could this contradiction be resolved? One way would be to target the probable speaker of both sentences, as in a Marxist argument concerning class. The first sentence mocking the conservative nature of biological cycles, as discussed with regard to gender roles, could have been uttered by a cultural critic. The second sentence follows the expansive logic of capitalist expansion: more nature is about to be manufactured. However, to solve the contradiction in terms of class interests does not seem convincing for this text for two reasons. First, it would rebut Jelinek's description of her own prose of being political in a formal way rather than in terms of content. Second, we should nor forget that this text, albeit referring to nature throughout, deals with wilderness. For the second reason, one could hardly expect a coherent solution for the adnate usages of nature. In the words of the text: "Pläne kennt die Natur so gut wie gar nicht."[24] Therefore, it seems

23 Jelinek, *Wildnis*, 56.
24 Jelinek, *Wildnis*, 82.

appropriate to assume that Jelinek's text emulates an ecological system of natures, a form of discourse revolving around different usages of nature targeted to show their political deficiencies by means of a formal manipulation of language. Although one might associate nature with a cyclical model of time, the usages within the text go beyond this cyclical model. For example, parts of this discourse contain economic planning, as we have seen in the sentence where nature is about to reach its "Endform," whereas other parts corroborate an indifferent stance towards time, supporting a perpetuation of the present, as discussed with regard to gender roles. In addition to future and present, "nature" also deals with the past with reference to a biological concept of decomposition. As for decay, the text rephrases the circle of life again: "Wir fangen noch einmal von vorne an, daher war nichts."[25] The regenerative power of nature was admired by many writers. Most famously, Walter Benjamin called this perspective an outstanding quality of the storyteller, being able to distance himself from current events, letting decades pass in a paragraph, looking at life's transient core of becoming and vanishing with the godlike eye of natural history. In Jelinek's book, nature's ability to decompose all of its parts resembles an uncanny quality as nature poses an antipode to history and memory. In her works, the disregarded past would become more relevant in later works. For instance, when zombies inhabit the Austrian soil in *Die Kinder der Toten* to haunt the untroubled, the ignorant, and some other *steirische Tiroler*. In *Oh Wildnis, oh Schutz vor ihr*, however, Jelinek alludes to the history of the lower class and the stories of unspecified strangers whom nobody cares about as nature deal with them.

> Die Helfer treiben dicht unter der Oberfläche der Natur dahin, träge spielen die Haare, eifrig schöpfen die Fische Nahrung aus den Augenhöhlen. Die Namen dieser Strauchexoten, Eigentum des Oberförsters, kennen sie alle nicht.[26]

> [Right below nature's surface, the helpers are drifting, their hair wafts inertly, avid fish scoop food out of the eyeholes. The names of these exotic bushes, property of the head forester, nobody knows them.]

25 Jelinek, *Wildnis*, 12.
26 Jelinek, *Wildnis*, 27.

Fish feed on the eyes of unknown helpers whose perspective will never be acknowledged as the owner of this land seems uninterested in revealing it. Here, the analogy to Jelinek's present political situation is obvious: no one is willing to reveal the offenders, even if they are sworn in as president. Of course, the literal paraphrase does not capture the form of these sentences. Apparently, nature and work are synthesized in the first sentence since fishes decompose the helpers. However, the description of the fish draws attention to the way in which they eat, "eifrig," pointing at the humanization of nature attributed to fish as to workers. Even when they are dead they will still be associated with diligence. In the second sentence, the helpers are referred to as "Strauchexoten," a common term for plants one finds in living rooms of the middle class. Combining a class distinction with exotic allusions to imperialism forces potential readers to identify themselves with the helpers while possibly owning these plants. Consequently, ownership stands as parenthesis in the middle of the second sentence, suggestive and disconnected from the content of these sentences, inciting readers to draw their own conclusions about the unwritten history of ownership and exploitation. The ecological system is thus not only concerned with cyclical models of time. It includes planning as well as memory, the future as well as the past. What all time-related aspects have in common, is the lack of a future horizon different from the biased present or the disregarded past as even the future threatens to reinforce the capitalist meaning of nature.[27]

Conclusion

I argued that *Oh Wildnis, oh Schutz vor ihr* reacts to the growth of the ecological movement by discussing a wider range of usages of "nature" than the one limited to the preservation of the wetlands. Jelinek's text problematizes "nature" on different levels. It polemically blurs the distinction

27 Ernst Bloch, *Das Prinzip Hoffnung* (Frankfurt/Main: Suhrkamp, 1985), 1628.

between nature and culture, highlights the discrepancy between nature as romantic topos and nature as critical concept with respect to economy and gender, and it discusses the ramifications of "nature" with regard to future, present, and past.

Overall, Jelinek's usage of nature is analogous with the unusual alliance seen in the press conference of animals, insofar as conservative values stand next to ecological goals and economic endeavors. Jelinek thereby calls attention to how the different positions within the ecological movement would lead to a political standstill, for they do not account for a clear policy with regard to nature, as nature turns out to be a wilderness of meanings. The classical illustration of a political standstill is depicted on the shield of Achilles in the eighteenth book of Homer's *Iliad*. His shield contains two depictions of the city. The city at peace is marked by rituals and sacrifice whereas the city at war is characterized by battles at the city gates. The task of politics is to be "polemical" because it provides the means to decide whether or not to engage in warfare. The city at war is characterized by a linear model of time, one event following the next one as recorded by history, while the city at peace is described in terms of a circular model of time, integrating people into the community by means of marriage and rituals.[28] According to the classical tradition, the circular model is part of nature, as it resorts to biological processes. Considering the opposition between war and peace, Jelinek's polemical text assesses the current political situation in Austria as one-sided because lacks policies in order to move forward in a certain direction, being unable to move from the circular to the linear model.

The assemblage of ecologists unites sexism and exploitation under the heading of an ever-continuing nature consolidating the deficits of past and present into a similarly gloomy future. To point out the lack of agenda is the political impetus of Jelinek's critique, yet the objective of her critique is not a concrete political goal, that is, she is not pleading for clear plan, be it a war or another agenda.

28 I am indebted to John Hamilton for pointing me to this reference. Homer, *The Iliad*, trans. Robert Fagles (New York: Penguin Classics, 1998), 467–488.

Throughout this text, my interpretation tried to make the critique explicit which was contained in the curt sentences of this text. However, it is not just the propositional form I exposed what this text aims to convey. This would reduce this text to an argument made in public: "Das öffentliche Sprechen über etwas ändert gar nichts. Die wollen einfach nichts lernen."[29] Jelinek seems cynical in this regard. Be that as it may, the emetic cue of this prose was to make readers vomit and only a literary text can do that by going through so many different usages of nature, ranging from natural resources over creaturely deterioration to floral sexism. Therefore, the positive goal of undoing the language derived from flora and fauna is to renew the political language as regards nature, for every usage Jelinek exposes in her text adds a caustic connotation to the existing language. The fact that Jelinek's writings polarize opinions seems to validate the quality and possible impact of her texts.[30] In this respect, her text conveys a political critique of the ecological movement by means of form. On the other hand, her text does not state what kind of policy should actually be followed given the political situation of the 1980s. As a writer, Jelinek finds a third way between explicit political critique and perfunctory aesthetic innovation. She creates innovative forms of writing in order to criticize the political present without giving way to the hubris a writer could immediately impact politics, refraining from a clear message.[31]

From a personal point of view, Jelinek's formal critique could be seen as frustrating endeavor because new usage of language could only affect readers who sympathize with her political positions anyway. The text gives an answer to the question what keeps Jelinek writing in spite of the possibility of political disappointment.

> Aus dem erschöpften Kessel der Vergleiche wird etwas in einen Teller gehoben. Dieser Abfallkübel (Sprache) versagt niemals, und zwar wegen Humors.[32]

29 Jelinek, *Wildnis*, 79.

30 There is but one balanced review of this book, written by Ulrich Weinzierl for the *Frankfurter Allgemeine Zeitung* on November 2, 1985.

31 Wolfgang Hackl, "Von der Sehnsuchtslandschaft zur Erinnerungslandschaft: Die Alpen im literarischen Landschaftsdiskus," in *Tourismus Journal* 8 (2004), 133–146.

32 Jelinek, *Wildnis*, 58.

> [From the jaded kettle of comparisons we put something into a plate. This waste bin (language) never fails to work, because of humor.]

Here the vanguard's dream to replace existing language reveals its own joys. Humor serves as fertilizer for the bittersweet growth of this text, whose wilderness aims to poison all denotations of nature. In retrospect, Jelinek's fear that the ecological movement would conflate the political Left did not turn out to be a real threat. After all, the Green Party in Austria became another left-liberal party, a green addition to the Social-democrats. Especially in the western part of the country, Tyrol and Vorarlberg, the Green Party would grow stronger over the next decades, as the existing conservatism was susceptible to a party which aimed to preserve the Green homeland rather than one standing for workers' rights. It is difficult to evaluate whether the way in which we speak about nature has actually changed. What I tried to show is that this text can change our consciousness for language, and one might enjoy being changed.

ELI RUBIN

The Greens, the Left, and the GDR: A Critical Reassessment

By the time the GDR collapsed, it was one of the most polluted territories in Europe. Over 9,000 lakes were biologically dead.[1] The groundwater near the Bitterfeld chemical factory was so polluted it had a pH between battery acid and vinegar.[2] Sulfur dioxide and other particulates spewed from aging smokestacks, choking cities with smog, and showering acid rain down on cities, forests, and crops. Rivers ran with contamination, and even the Palace of the Republic, a marvel of socialist-modern architecture, turned out to be packed full of asbestos. As the GDR began to dissolve over 1990, the extent of the pollution in East Germany became a revelation to those in the West, and even to many East Germans.

1990 was a year of many revelations in East Germany. The extent of the GDR state's human rights abuses also came to light at the same time, especially the massive domestic surveillance and persecution perpetrated by the Stasi. The extent of the state's *Mißwirtschaft* [financial mismanagement] and indebtedness also was revealed once the GDR's books were opened. All of these disclosures unfolded in connection with each other over the course of the *Wende*. The same was true throughout Eastern Europe, as state socialism crumbled and both the extent of the environmental contamination and the depth of human rights abuses came to light. It was Vaclav Havel who best helped frame this in his first inauguration speech,

1 David Blackborn, *Conquest of Nature: Water, Landscape, and the Making of Modern Germany* (New York: Norton, 2006), 335.

2 Blackborn, *Conquest*, 339. Also see Sandra Chaney, "A Chemical Landscape Transformed: Bitterfeld, Germany Since 1980," *Global Environment* 10 (2017), 137–167.

given in January 1990 at the height of the Velvet Revolution. Describing the pollution spewing from the enormous Slovnaft chemical factory, he bemoaned the fact that Czechoslovakia had "polluted the soil, rivers and forests bequeathed to us by our ancestors" and then linked that to the fact that "the worst thing is that we live in a *contaminated moral environment*" (my emphasis).[3] Environmental pollution was not coincidental to State Socialism, for Havel – the two flowed from the same root. State socialism caused pollution because it cared for neither the well being of nature nor the individual; its terrible treatment of the environment was therefore one of the factors that caused its downfall.[4]

Havel's perceived link between environmental pollution and state socialism has been amplified and redeployed in both popular and academic discourse over the last three decades. Environmental historian Frank Uekötter has written that there is a "disaster school" of Eastern European history, which focuses on the environmental degradation in the USSR and Eastern European history.[5] A prominent theme of this school is that state socialism, especially its Stalinist variant, instrumentalized nature. It saw nature as a resource to be used up in the endless quest for industrial growth, and also as a problem that needed to be corrected.[6] The

3 Havel, *The Art of the Impossible: Politics as Morality in Practice: Speeches and Writings, 1990–1996* (New York: Knopf, 1997), 3–4.

4 Eagle Glassheim makes this point, also about communist Czechoslovakia, in his *Cleansing the Czechoslovak Borderlands: Migration, Environment, and Health in the Former Sudentenland* (Pittsburgh, PA: University of Pittsburgh Press, 2016), 11, 121, 171, and passim. Glassheim argues convincingly that ecology came more and more to be at the center of Havel's moral critique of communist "materialism." Also see Adam Fagan, *Environment and Democracy in the Czech Republic: The Environmental Movement in the Transition Process* (Northhampton, MA: Edward Elgar, 2004).

5 Uekötter, *The Greenest Nation? A New History of German Environmentalism* (Cambridge, MA: MIT Press, 2014), 14. Uekötter mentions that some scholars have begun to question this "disaster school" as oversimplified, but mentions only one, an article by Julia Obertreis, "Der 'Angriff auf die Wüste' in Zentralasien: Zur Umweltgeschichte der Sowjetunion," *Osteuropa* 58, 4–5 (2008): 37–56.

6 See Doubravka Olsakova, ed., *In the Name of the Great Work: Stalin's Plan for the Transformation of Nature and its Impact in Eastern Europe* (New York: Berghahn, 2016).

instrumentalization and "improvement" of forests and rivers, furthermore, was connected to the instrumentalization and "improvement" of humans by a gargantuan state, a process portrayed in negative terms. A recent work on 1948 Stalin Plan for the Transformation of Nature, which was a plan to straighten flood prone rivers, build massive windbreaks to halt erosion of soil on the Steppes, open up mining, and generally harness the vast natural resources of the USSR, uses the terms like "totalitarian," "Kafkaesque" and "conspiracy" to describe this subject.[7]

Another book on the East German forest written in 2005, published by Yale University Press, claimed that the "constant pollution" was an outward sign of the "Marxist-Leninist Ponzi scheme" that characterized the GDR and accused the GDR of "Marxist-Leninist hypermaterialism" which failed to put ecology and economy in "on equal footing." The East German forest's "decline and economic ruin" was directly connected to the East German state's deserved "dissolution into oblivion."[8] Other analysts and commentators in the 1990s confirmed that it was communism *per se* that was responsible for destroying the environment, because, allegedly, in Marxist ideology "natural resources are free and have no intrinsic value" and because absent a profit motive, such as exists in capitalism, socialist planned economies had no in-built restraints which could force them to be more efficient in their use of natural resources.[9]

Not only was the environmental pollution connected inherently to a system that was undemocratic and abused human rights, but also, for that very reason, the fight against pollution naturally led to a bigger fight for freedom and a *Rechtsstaat*.[10] Thus, the grassroots movements that

7 This is Olsakova, ed., *In the Name.*

8 Arvid Nelson, *Cold War Ecology: Forests, Farms, and People in the East German Landscape 1945–1989* (New Haven, CT: Yale University Press, 2005), xii–xiv.

9 Dominick, "Capitalism, Communism, and Environmental Protection," 312.

10 See Obertreis, "Von der Naturbeherrschung zum Ökozid? Akuelle Fragen einer Umweltzeitgeschichte Ost- und Ostmitteleuropas," *Zeithistorische Forschungen* 1/12, accessed November 26, 2017, <http://www.zeithistorische-forschungen.de/site/40209222/default.aspx#pgfId-1037045> and her discussion of the "master narrative" of Eastern European/Soviet socialist nations as the "great ecological sinners" in the context of the Cold War.

sprang up to protest this pollution, and faced repression from the regime, became the "nail in the coffin" of the GDR, according to historian Ilko-Sascha Kowalczuk.[11] Historian Tobias Huff dates the true "beginning of the end" of the GDR, and perhaps therefore the Cold War itself, to the infamous Stasi raid on the "environmental library" kept by dissidents in the basement of the Zion Church in East Berlin in November of 1987, so central was the role of environmental protest in the drama that led to the *Wende*, in his view.[12]

David Blackbourn's 2006 *Conquest of Nature* goes even further, portraying the GDR as the heir of a promethean tradition in German history – a long standing tendency, dating to the Enlightenment, of the state and its allies to embark on projects that dominated or "conquered" nature. In fact, as Blackbourn describes it, the GDR, and the Soviet communism that profoundly shaped it, represented Enlightenment prometheanism on steroids – it was a "heroic Soviet model" of planning which "aimed to correct nature's mistakes."[13] The GDR was, according to Blackbourn, "in thrall to Soviet gigantomania" in its pursuit of large industrial projects such as Eisenhüttenstadt (an enormous steel plant and factory town built in the 1950s on the GDR-Polish border, originally called Stalinstadt), and in the GDR "nature was ... there to be beaten into submission."[14] Blackbourn ties this back to the fact that Marxism was, in its essence, a promethean ideology and that in Marxist ideology, the end of capitalism "would end the exploitation of man by man, leaving humankind free to exploit nature."[15] Key to Blackbourn's narrative is that "the GDR reached an environmental turning point and failed to turn,"[16] a turn which the Federal Republic

11 Kowalczuk, *Endspiel. Die Revolution von 1989 in der DDR* (Munich: Beck, 2009), 126–127.

12 Huff, *Natur und Industrie im Sozialismus: Eine Umweltgeschichte der DDR* (Göttingen: Vandenhoeck & Ruprecht, 2015), 399.

13 Blackbourn, *Conquest*, 336; see also Olsakova, ed., *In the Name of the Great Work*, especially Paul Josephson, "Introduction: The Stalin Plan for the Transformation of Nature, and the East European Experience," 1–42.

14 Blackbourn, *Conquest*, 336–37.

15 Blackbourn, *Conquest*, 335.

16 Blackbourn, *Conquest*, 335.

did make, once Green politics became a major force in the 1970s in West German politics and culture, and the FRG cleaned up its environment through regulations on industry, similar to other western nations.

Blackbourn's implied teleology has become widely accepted: there is a standard trajectory that nations follow in the course of their development, from an industrial to a post-industrial era, and learning to be "green" and leave behind their prometheanism is part of that maturation. Frank Uekötter, similarly, has surmised nations get less "red" and more "green" as they "mature" – what he calls the reverse tomato theory.[17] If they get "stuck" in the industrial, promethean (or "red") phase, and continue to pollute their environment, the laws of history and ecology will punish them with political collapse, and this is just what happened to East Germany and the other socialist nations.

Like all narratives, this is a constructed narrative. And like all narratives, it was constructed within a context – in this case, the Cold War, and specifically, the Cold War competition between the FRG and the GDR. This narrative was, we might say, akin to a "Green Legend," similar to the "Black Legend," which was used by the British in their colonial competition with the Spanish – a legend built, to be sure, on many truths, but one created under the conditions of global antagonism which also obscured as much as it revealed. This essay is an initial exploration into how this "Green Legend" was constructed – not by any one person or group of people, but, as these things go, by the interaction of multiple dynamic forces. It argues that the development of environmentalism, the Green Party and the environmental regulation which did propel Germany into being such a "Green" nation, was the result of internal divisions on the Left in West Germany, as well as a longstanding aversion to Soviet state socialism, especially among the many 68ers who joined the Green movement. It argues further that Green politics, once it became mainstream in the FRG, became weaponized in the German-German context, causing an overreaction in the GDR, leading to a kind of self-fulfilling prophecy in which civil rights abuses compounded and became linked

17 Uekötter, *Greenest Nation*, 101.

to environmental contamination. Finally, this essay is in part a response to Frank Uekötter's call for environmental historians to put their narratives in a broader historiographical context – in this case, the Cold War.[18] Ultimately, it argues, that when we critically deconstruct the history of this narrative for what it is – a Cold War narrative – we can come to a better understanding of the contradictions and unresolved blind spots of modern environmental issues.

***

Environmentalism is often assumed to be a movement of the "Left," but in Germany, a closer examination shows that what we think of as "Green" politics has been shaped by influences from across the ideological spectrum – before the 1970s, in particular, what we call now "environmentalism" was really known as "conservation," and conservation groups like the *Deutscher Bund Heimatschutz* at times verged into right-wing mystical views of nature.[19] Jost Hermand argues that there was a strong trend of "green utopianism" which embodied a fundamental critique of modernity, progress, technology and growth present in German culture since the Enlightenment.[20] For Hermand, the modern environmental or "Green" movement was not as new nor as progressive as it may have seemed – he dates the birth of the Green movement to the publication of *The Limits to Growth* commissioned by the elitist Club of Rome in 1972.[21] *The Limits to Growth* was a critique of "growth" but it also rested on neo-Malthusian assumptions that, when drawn out to their logical conclusions, were deeply troubling: the industrialized world had achieved a certain living standard,

18 Uekötter, *Greenest Nation*, 102.

19 Franz-Josef Brüggemeier, Mark Cioc and Thomas Zeller, "Introduction" in Brüggemeier, Cioc and Zeller, eds., *How Green were the Nazis? Nature, Environment, and Nation in the Third Reich* (Athens: Ohio University Press, 2005), 5. For an excellent accounting of conservation in postwar West Germany, see Sandra Chaney, *The Nature of the Miracle Years: Conservation in West Germany, 1945–75* (New York: Berghahn, 2008).

20 Hermand, *Grüne Utopien in Deutschland. Zur Geschichte des ökologischen Bewußtseins* (Frankfurt/Main: Fischer, 1991).

21 Hermand, *Grüne Utopien*, 17 and 134–35.

through unbridled growth and industrialization, and now that same process was to be denied to the rest of the globe, even though much of the developing world had been exploited through colonialism to support the very growth that lined the pockets of the members of the Club of Rome in the first place. Other scholars see similar Malthusian forces behind the rise of modern environmentalism, including Thomas Robertson, who argues that even before *The Limits to Growth*, Paul Ehrlich's 1968 *The Population Bomb* and Garrit Hardin's 1968 essay "The Tragedy of the Commons" set in motion the combination of forces of neo-Malthusianism and ecological justice which created the environmental movement in its modern and politically potent form.[22]

In West Germany, the Green movement traced its roots back to the anti-nuclear protests of the 1970s, and the foundational event of this movement were the protests and occupations at the proposed nuclear reactor site in Wyhl, near Freiburg, in 1975, an event involving tens of thousands of protestors and hundreds of police, which became a "national myth" according to Uekötter.[23] Jutta Ditfurth, one of the activists at Wyhl, similarly called it one of the "main roots" of what was to become the Green Party.[24] But Wyhl is significant not just because it was foundational to the Green Party, but also because of the remarkable ideological *heterogeneity* of the participants there. Alongside "leftists," like Ditfurth, fought local residents, farmers and vintners, students, Christians, and (even more so at other protest sites such as Brokdorf), right-wing "eco-fascists," as Ditfurth called them.[25] A loose coalition of environmental groups emerged after Wyhl, able to mobilize similarly enormous numbers of protestors – ranging from the 10,000s to the 100,000s[26] to stop other projects, including

22 Robertson, *The Malthusian Moment: Global Population Growth and the Birth of American Environmentalism* (New Brunswick, NJ: Rutgers University Press, 2012), 151–154.
23 Uekötter, *Greenest Nation*, 95.
24 Ditfurth, *Das waren die Grünen. Abschied von einer Hoffnung* (Munich: Econ, 2000), 55.
25 Ditfurth, *Das waren die Grünen*, 55 and Uekötter, *Greenest Nation*, 92.
26 Dieter Rucht and Jochen Roose, "Germany" in *Environmental Protest in Western Europe*, ed. Christopher Rootes (New York: Oxford, 2003), 81–82.

other nuclear power plants, chemical plants, airports, etc., making it one of the largest grassroots mass movements in history.[27]

Part of what made this movement able to mobilize such enormous numbers of citizens, despite having no central organization and having only been around for a few years, was its ability to appeal to people from so many different ideological backgrounds – from the left, the right, and people who fit nowhere neatly on the traditional spectrum. But how was it able to do that? In short: a critique of "growth." The magic formula for drawing left, right, and "other" together under one banner was a critique of the modern, bureaucratic, technological society which seemed to care neither about the environment nor the individual, but rather was driven inexorably by the logic of ever-increasing growth, both of the economy and of the state. For this reason, state-sanctioned and state-protected nuclear projects like Wyhl crystallized the image of everything that activists like Ditfurth feared most deeply – as she put it, she was opposed to the idea of the "nuclear state" itself,[28] by which she meant a state large enough, technologically advanced enough, and blind enough to nature and humans that it would embark on splitting atoms to satiate the endless demand for more energy, was a state that could quickly produce both nuclear weapons and crush individual rights – in short, a Moloch able to crush atoms and people alike.[29]

Activists like Ditfurth considered themselves part of the "undogmatic" Left, to distinguish themselves from the more established, etatist, bureaucratic Left which existed in the USSR and Eastern Europe, what some might have called the "Old Left," as distinguished from what was becoming the "New Left." This Old Left, in the view of the "undogmatic" or "New" Left, was guilty of embracing endless economic growth and technological progress as well as the bureaucratic,

27 Ditfurth, *Das waren die Grünen*, 65.

28 Ditfurth, *Das waren die Grünen*, 55.

29 Uekötter, *Greenest Nation*, 93. The image of the Moloch was used by anti-Nuke and environmental activists frequently in the 1970s, including protestors against the VEBA chemical factory in Rheinberg in 1971.

technocratic state that pushed this "progress" at the expense of nature and individuals.[30]

This combined critique – both of the state as a Moloch existing only to perpetuate its own power, and of the pursuit of relentless economic and technological growth, which was often associated with the modern bureaucratic state – became a core feature of the "New Left" movement, and can be traced in part to the group of theorists known as the Frankfurt School. Frankfurt School theorists produced a left-wing cultural critique that diverged from, as many at the time understood it, classic Marxist materialism which saw class conflict as the engine of history. After 1945, Martin Jay writes, class conflict was replaced by the "larger conflict between men and nature" as the motor of history.[31] The main fight was not socialism vs. capitalism any more – it was modernity vs nature and the individual, and for many, socialism *and* capitalism were the oppressors. This was important – it paved the way for a Green equivalency of socialism and capitalism, and a critique of both, which would come into play later during the rise of the western Greens in the 1980s.

Frankfurt School theorist Alfred Schmidt, in particular, was responsible for creating a critique of Marx's theory as *dualistic* – meaning it set up an inherent conflict between the human subject and the objective world around it, including nature, which could only be resolved through the subject transcending the object, by "mastering" it – prometheanism, in other words.[32] Equally as influential was Carl Amery, whose 1976 *Natur als Politik* sought to build upon this foundation to create a new "ecological

30 Ditfurth, *Krieg, Atom, Armut. Was sie reden, Was sie tun: Die Grünen* (Berlin: Rotbuch, 2011), 42–43.

31 Neil Smith, *Uneven Development: Nature, Capital and the Production of Space* (New York: Blackwell, 1984), 29.

32 Smith, *Uneven Development*, 26; see also Alfred Schmidt, *The Concept of Nature in Marx*, trans. Ben Fowkes (London: NLB, 1971) 38–39. Theodor Adorno also wrote about the problem of humans "dominating" nature in in his writings on the concept of culture (which shares the same root as "cultivate" and "agriculture," a process of mastering nature for human needs); see his transcribed remarks "Kultur und Culture" [Kultur and Culture], in *Hessische Hochschulwochen für staatswissenschaftliche Fortbildung* (Bad Homburg: Max Gehlen, 1958), 246–59.

materialism" to transcend Marxian political economy, in which natural history would replace social history, and in which the driving contradiction of history is human exploitation of nature, not human exploitation of humans.[33] At the same time, the French Marxist Louis Althusser also argued for a transition away from class conflict as the main struggle, suggesting that it was state versus the individual that represented the core conflict.[34]

The Frankfurt School, and the broader "New Left" to which it was connected helped fundamentally shape the "1968ers" – the rebels and revolutionaries who participated in the May revolts in France or in the Extraparliamentary Opposition (APO) in West Germany, many of whom went on to become leaders of the environmental movement and the Green Party. The critique of the technocratic state and its insistence on "growth" was one of the key ideas that the 68ers inherited from the New Left, and it paved the way for a later Green critique of East Germany in the context of the Cold War. So for example, Danny Cohn-Bendit, the unofficial leader of the May 1968 uprisings in France, was opposed to the French Communist Party (PCF), which he and his comrades saw as sclerotic, bureaucratic, and only interested in "growth," *and* to Soviet Communism, which seemed to care only about building more factories, plants, and succeeding in more technological breakthroughs. Cohn-Bendit lumped both the PCF and Soviet State Communism together, calling this "obsolete communism" (he used the term "*la maladie senile du communisme*" in French) and proclaiming in 1968 that in refusing to back the May uprising in Paris, the PCF "simply acted in defense of its bureaucratic interests and of the bureaucratic interests of the USSR as a state."[35] Later, as a successful Green politician, Cohn-Bendit would proclaim himself "the most militant anti-Communist in the Federal Republic."[36]

A close comrade of Cohn-Bendit's in West Germany was Joschka Fischer, who, along with Cohn-Bendit, edited the magazine *Pflasterstrand*.

33 Andrei Markovits and Philip Gorski, *The German Left: Red, Green and Beyond* (New York: Oxford, 1993), 127.

34 Smith, *Uneven Development*, 30.

35 Daniel and Gabriel Cohn-Bendit, *Obsolete Communism: The Left-Wing Alternative*, trans. Arnold Pomerans (London: André Deutsch Ltd., 1968), 18.

36 Ditfurth, *Das waren die Grünen*, 222.

Pflasterstrand was a central organ for the so-called "*Spontis*," ex-68ers who rejected not only the modern, western, managerial corporate state, but also the modern eastern bureaucratic Party state.[37] Fischer, moreover, was a disciple of one of Adorno's protégés at the Frankfurt School, Hans-Jürgen Krahl.[38] The *Spontis*, led by Fischer and Cohn-Bendit, entered into the Green Party and effectively overtook the Green Party in Frankfurt, one of the most important centers of Green activism in West Germany.[39] The Frankfurt School's rejection of pure class conflict in favor of man vs nature dualism dovetailed well with the 68er-*Spontis* rejection of both the modern corporate state and the Soviet/Eastern European communist state, which emphasized technology and heavy industry; and both of these trends found a new political home in the vibrant Green movement in Frankfurt, which was organized largely out of anti-nuclear activists. Other formerly left-wing activists and intellectuals, like Hans-Magnus Enzensberger, Carl Amery, Heinrich Böll and Günter Wallraff went green as well, shifting the focus of their condemnation to the "human domination of nature"[40] and away from the human domination of other humans.

The 68er/*Spontis*' critique of "growth" and the "state" allowed them to navigate safely the dangerous political shoals of domestic politics in Germany during the Cold War: it allowed them to remain on the Left while distancing themselves from both the Soviet Union and its allies, and at the same time not embracing the corporate capitalism that held such sway over mainstream West German discourse. Uekötter argues that Green politics opened up a pathway for ambitious Leftists to actually have meaningful careers in the Federal Republic:

> It was in fact quite an ideological feat to move from a Marxist vocabulary to an ecological one: after all, the exploitation of the proletariat was something rather different from the exploitation of nature. However, dogmatic precision did not reign supreme for a left that had suffered a long string of defeats since the late 1960s. In a way, the left tried almost everything else to gain popularity, only to strike upon

37 Ditfurth, *Das waren die Grünen*, 86–88.
38 Ditfurth, *Das waren die Grünen*, 87.
39 Ditfurth, *Das waren die Grünen*, 88
40 Blackborn, *Conquest*, 332.

> ecology at the very end. The Left's penchant for green causes always had an air of political opportunism. For an embattled left, ecology was tantamount to a last hope ... it was good enough for a leftist that the masses were finally listening to them. Why bother that they were revolting for the wrong reasons? From the point of view of a German Marxist, going environmental meant becoming popular.[41]

Uekötter's comments are infused with at least a touch of cynicism, which is important:

For those original environmentalists, like Ditfurth, who saw themselves as the "real activists," the entrance of self-styled revolutionaries like Cohn-Bendit and Fischer was more about the "arrival of the careerists."[42] Indeed, both Cohn-Bendit and Fischer did go on to have very successful political careers as Green Party politicians, with Fischer eventually becoming Foreign Minister under Gerhard Schröder and a major figure on the world stage, and Cohn-Bendit an important figure in the European Parliament. It showed that a critique of growth – of human exploitation of nature, instead of human exploitation of other humans – was the move that could open doors to political power for former Marxists in the Federal Republic.

The critique of growth was so powerful precisely because it could attract groups from the Right as well as the Left. Present at many of the gatherings, protests, political meetings, etc. that formed the rising green tide in 1970s West Germany were far-right groups as well, including the *Weltbund zum Schutz des Lebens* [World Union for the Protection of Life] (WSL), a *völkische* group founded by former Nazi official Werner Georg Haverbeck in 1958.[43] The WSL and other similar groups held, at the center of their ideology, the belief that out of control industrial and technological growth threatened nature, resources, and especially the nature and resources of the white European race – a threat on par, as they saw it (and still see it) with non-white immigration to Europe. In particular, they continued the quasi-mystical worship of nature that come out of racist *völkisch* thought, and it was through this right-wing ideological pathway that they arrived at modern environmentalism. In fact, the surprising reality is that it was

41 Uekötter, *Greenest Nation*, 109–110.
42 Ditfurth, *Das waren die Grünen*, 86
43 Diehl, "German Ultra-Right," 44.

these "eco-fascists"[44] who were among the trailblazers of modern Green activism, being among the very first anti-nuclear protestors.[45]

At the center of these "eco-fascists" stood Herbert Gruhl, a CDU politician and ideologue of right-wing environmentalism. Gruhl rose to prominence as the author of a 1975 book titled *Ein Planet wird geplündert* [*A Planet Is Being Plundered*], which warned of a "human flood" and a "human landslide" (*Menschenlawine*) of black and brown people forced from their lands, and into Europe, due to overpopulation.[46] According to Gruhl, incessant and unplanned industrial progress, economic growth, and technological innovation, core facets of both capitalism and communism, were precipitating a massive demographic crisis, mainly in the developing world, where a population explosion was outstripping both the carrying capacity of the land and therefore the ability of the economy to absorb the excess population. As he wrote, "out of control procreation of the Third World" was "throwing 200,000 people on to the job market daily" which was going to lead shortly to "unimaginable revolutions."[47] This was not so different from the conclusions of the *Limits to Growth*, only with more explicit racist tones. For Gruhl and his followers, part of what made white European culture "triumphant" and superior was that it had learned to manage its population to be in harmony with its natural resources, including learning to lower its own birthrate – a fascinating about-face from the earlier concerns about low white birthrates voiced by nationalists and racists in Europe, and another example of how the green umbrella opened up new pathways for old ideologies.[48] And, to solve such problems, liberal democracy was not up to the task – it would take a strong state, possibly

44 Diehl, "'Ecology' and the Modernization of Fascism in the German Ultra-right" in Diel and Peter Staudenmeier, *Ecofascism: Lessons from the German Experience* (San Francisco, CA: AK Press, 1995), 32–67.

45 Joachim Radkau, *Natur und Macht: Eine Weltgeschichte der Umwelt* (Munich: Beck, 2000), 304.

46 Ditfurth, *Feuer*, 235.

47 Gruhl, "Die Menschheit ist am Ende" *Der Spiegel* 13, March 23, 992; Quoted in Ditfurth, *Feuer*, 233.

48 Ditfurth, *Feuer*, 237. See also Janet Biehl and Peter Staudenmeier, *Ecofascism: Lessons from the German Experience* (San Francisco, CA: AK Press, 1995).

a kind of eco-dictatorship to hold back the flood of refugees and preserve the pristine German and European lands and resources.[49]

Opposition to growth was clearly not just a leftist issue, then – it could be found at opposed ends of the ideological spectrum, and all along in between. Likewise, opposition to growth was not just a western movement, but could be found on both sides of the "iron curtain." Perhaps no one illustrates the ability of anti-growth environmentalism to transcend the east-west divide more than Rudolf Bahro, one of the founders of the West German Green Party *and also* an East German dissident. Bahro had been arrested, jailed, and then banished from the GDR for having written *Die Alternative: Zur Kritik des real existierenden Sozialismus* ["The Alternative: Towards a Critique of Real Existing Socialism," translated into English as *The Alternative in Eastern Europe*], published in Cologne in 1977. Inspired by his disgust of the Warsaw Pact's crushing of the Prague Spring in 1968,[50] *The Alternative* was an attempt to reimagine a socialist planned economy that did not orbit so much around Fordism, which expressed itself in Plan quotas, phrases like *Tonnenideologie* ["ideology of tons," a common phrase used in East Germany to denote a focus on industrial production] and *rasantes Wachstum* [fast-paced growth].[51] Nor was Bahro alone; other East German dissidents, such as Wolfgang Harich and the scientist Robert Havemann published works questioning the idea of "growth" in communism, and arguing that unchecked growth and unbounded optimism in technology and industry were existential problems for both communism and capitalism.[52] Harich wrote in his 1975 *Kommunismus ohne Wachstum?* [*Communism without Growth?*] that at least 70 percent of the products made by industry (in both the West and the East) were superfluous, and

49 Ditfurth, *Feuer*, 237.

50 Bahro, *From Red to Green: Interviews with the New Left Review*, trans. Gus Fagan and Richard Hurst (New York: Verso, 1984), 62.

51 Bahro, *The Alternative in Eastern Europe*, Chapter 14 ("Cultural Revolution II") for a good summation of his ideas for reforming state socialism at the time.

52 Havemann, *Morgen: Die Industriegesellschaft am Scheideweg: Kritik und reale Utopie* (Munich: Piper, 1980) and Harich, *Kommunismus ohne Wachstum? Babeuf und der Club of Rome* (Reinbek bei Hamburg: Rowohlt, 1975).

criticized Honecker and the SED for trying to match western norms of consumption, which were themselves wrong.[53] Havemann similarly called for an arcadian utopia which would be neither socialist nor capitalist, but something else: a quasi-deindustrialized society in which *the ecology* was the organizing principle, as opposed to *the economy*.[54] Havemann and Harich's work were important voices, but it was Bahro whose work was by far the most comprehensive and coherent, coming to form what Andrei Markovits and Philip Gorski claim was the defining theory or ideology of the Green movement.[55]

The thrust of Bahro's thought in *The Alternative* was that "real existing socialism," which was the cornerstone of the Honecker era in the GDR, was attempting to mimic the consumer society of the West, and it was pouring tremendous resources into industrial production, resources it did not have to do so – and that all of this deviated from Marx's and Engels' original ideas about the transition to communism. Though Bahro did not have access to most of the work of the Frankfurt School, he claimed to be indirectly influenced by them, and their critique of mainstream popular and consumer culture in the West, and the massive costs associated with them, were echoed in his critique of Eastern European socialist consumerism.[56] Understandably, then, *The Alternative* was very popular in West Germany, especially among the "undogmatic Left," the 68ers and many in the Green movement who longed for a critique of both western capitalism and Eastern European state socialism.

Bahro was incarcerated in the GDR in 1977 and banished in 1978, being released to the West. By the time he arrived in West Germany in 1978 he was already a *cause célèbre* among extra-parliamentary Leftists and Greens, including Fritjof Capra, one of the leaders of the Green movement, who rightly discerned in Bahro's work the ideas of a "hidden Green."[57] As

53 Hermand, *Grüne Utopien*, 144.
54 Hermand, *Grüne Utopien*, 145.
55 Markovits and Gorski, *The German Left*, 126.
56 Bahro, *From Red to Green*, 60–61.
57 Charlene Spretnak and Fritjof Capra, *Green Politics* (Santa Fe, NM: Bear & Company, 1986), 25.

both a critic *and* a victim of a Moloch-like bureaucratic state which – seemingly – crushed both nature and individuals beneath it, Bahro was attractive to West German Greens as both a symbol and a leader, and quickly found his political home in the Green movement.[58] Here, he quickly began to distance himself from both capitalism *and* socialism. In a speech given at the founding congress of the Green Party in 1980, entitled "Why, as a Socialist, I'm Joining the Green Party" he claimed: "Without overcoming the ecology crisis, which puts in question the very existence of human civilization on this earth, the mere possibility of the socialist goal – the general emancipation of human beings, men and women – becomes an illusion."[59]

He went on to say:

> In order to become master of nature, humanity had to reproduce itself and its tools on an expanded scale. This was at one time a progressive necessity. But in order to remain in nature and to gain control over itself, humanity must stabilize itself in its relationship with nature ... Gorki once said of the difficulty in changing the world that the basic theme of all moans and groans on the part of the majority was; "don't stop us living as we are used to," and I would add: "Don't stop us expanding as we are used to." Our social life is organized in such a way, both here and in the other bloc, that even people who work with their hands are more interested in a better car than a single meal of the slum-dweller on the Southern half of the earth.[60]

> If what is going on now continues for a further twenty, thirty, forty or fifty years, then our Earth will be stripped bare. We shall be like a swarm of locusts who multiply and fall upon a certain territory, eat it bare and then all perish themselves.[61]

Though he did not share Gruhl's racist fear of a Third World *Menschenlawine* flooding into Europe, he believed the threat posed to the environment

58 EP Thompson, for example, called the *Alternative* "one of the few necessary, original, and truly significant contributions to the political thought of Europe in the post-War years ... essential reading." See Thompson, "Introduction" to Bahro, *Socialism and Survival: articles, essays and talks 1979–1982*, trans. David Fernbach (London: Heretic Books, 1982), 7.

59 Bahro, "Why, as a Socialist, I'm Joining the Green Party," in Bahro, *Socialism and Survival*, 57.

60 Bahro, "Ecology Crisis," 30–31.

61 Bahro, "Ecology Crisis," 39.

by economic growth transcended all other ideological divides, whether East-West or Right-Left. Thus it was that Bahro, a former SED functionary, came to argue that the Greens needed to accommodate eco-fascists like Gruhl,[62] writing at one point "I agree to a large extent with Herbert Gruhl's formulation: neither to the right nor to the left in the traditional sense, but ahead."[63] Bahro's pursuit of this transcendence led him down strange and eccentric paths, through mysticism, anthroposophism, theosophy, and New Ageism, a journey he shared with other former 68ers, including Rainer Langhahns, with whom he ended up living on a New-Age commune, and though he lost personal influence in the Green Party (in part due to his increasing eccentricity), his ideas continued to shape it: in particular, the idea of transcending Right and Left, became, according to some, an "unofficial slogan" of the Greens.[64]

It is not clear how much right-wing or *völkische* ideology Bahro absorbed, though he purportedly called once for an "eco-dictatorship" led by a "Green Adolf" as the only thing that could really put a stop to growth.[65] But it was clear that he had put leftism in his rear-view mirror for good, and his journey away from Marxism was through the Greens. Bahro came to believe that Marxism was an outmoded and rigid dogma which could not account for the fullness of life in a holistic sense, and which – though it claimed to be an account of dialectics – itself needed to be transcended dialectically, a process that was undoubtedly a *green* movement. "The Greens are to Marx and Marxism what Einstein was to Newton and Newtonian physics – in short, a qualitative transformation of

62 Bahro, "Ecology Crisis and the Socialist Idea," in Bahro, *Socialism and Survival*, 35.

63 Bahro, "Ecology Crisis,"35.

64 Roland Rolf and Detlef Murphy, "From Competing Factions to the Rise of the Realos,"., *The German Greens: Paradox Between Movement and Party*, eds. Margit Meyer and John Ely (Philadelphia, PA: Temple University Press, 1998, 49–71, 51.

65 See Peter Kratz, "Bahros grüne Adolf *Die 'Neue Rechte' an der Berliner Humboldt-Universität*'" accessed December 12, 2017, <http://www.bifff-berlin.de/Bahro.htm>; also see Ditfurth, *Feuer*, 300–302 – Ditfurth notes that Bahro at times claimed he never used these terms, and at times that he did but in an interview taped without his permission and taken out of context.

a worthwhile system whose time, however, is up," he wrote.[66] Ultimately, Bahro concluded that "no more leftist projects" were possible.[67]

The transition away from Marxism, through the gateway of anti-growth ideology, was mirrored by many 68ers. Even for those in the Greens who tried to retain their Marxist worldview, such as so-called "eco-socialists" like Rainer Trampert and Thomas Ebermann, there was an acceptance that the working class would have to embrace an end to "growth."[68] This brought them to loggerheads with both the western SPD and the USSR. The SPD's decision to align with mainstream politics and focus on "bread and butter" issues like jobs and therefore economic growth led it to be on the wrong side of many key controversies according to the Greens, including the Startbahn West airport extension in Frankfurt.[69] And for many Greens (including Bahro), it was the invasion of Czechoslovakia in 1968 that sparked their disillusionment with state socialism, and the USSR.[70]

Even the hard-left groups within the Green coalition, the so-called "K-Gruppen," were Maoist in orientation, and therefore ideologically opposed to the USSR. Even they came to admit, according to Fritjof Capra, that "socialist commandeering of the power structures in Eastern Europe had not transformed society in a positive way."[71] Of course, other wings of the Green coalition such as the eco-libertarians[72] and the eco-fascists, though small in number, were strongly opposed to state socialism of any kind. Anti-communism (at least anti-*Soviet* communism) was a unifying force for the tempestuous and heterogeneous coalition, and it became accepted wisdom among the Green leadership. Joschka Fischer argued in the mid-1980s that "capitalism has won, socialism has lost" and stated, with conviction, that any salvation of the economy had to come from "using

66 Spretnak and Capra, *Green Politics*, 25.

67 Markovits and Gorski, *The German Left*, 141.

68 Markovits and Gorski, *The German Left*, 147.

69 Thomas Scharf, "Red-Green Coalitions at the Local Level in Hesse" in *Greens in West Germany*, ed. Kolinsky, 162.

70 Markovits and Gorski, *The German Left*, 9.

71 Spretnak and Capra, *Green Politics*, 22.

72 Markovits and Gorski, *The German Left*, 145.

the forces of the market" and *not* the power of the state. Like Bahro, he proclaimed that there was no more basis for "left politics."[73] Petra Kelly, another leading Green activist and politician, reportedly "never considered herself part of the Left."[74] No less than Otto Schily, Gudrun Ensslin's attorney and founding member of the Greens, claimed in 1987 that the Greens were "definitely not a party of the Left" and had "essentially conservative values."[75] There were some exceptions, such as Hans-Magnus Enzensberger who criticized the *Limits to Growth* in his 1973 *Zur Kritik der politischen Ökologie* [Critique of Political Ecology] by arguing that the focus in the Green movement should remain capitalism, but he was, as Jost Hermand calls him, a "loner."[76] In reality, as Frank Uekötter has written, the environmentalism of the 1980s was focused on "middle-of-the-road topics" and not social justice, even though there were plenty of opportunities to highlight the links between environmental pollution and social discrimination, saying that "with a view to the socioeconomic reality in Germany, the ecological 1980s always had a whiff of escapism."[77]

The confluence of anti-communism (especially Soviet and Eastern European communism) with environmentalism crystallized perhaps most clearly in the figures of Wilhelm Knabe and his son, Hubertus. Wilhelm Knabe was an ecologist and forestry official in Tharandt, East Germany, who specialized in researching the effects of smokestack emissions on forests. In 1959, Wilhelm left the GDR for the FRG, where he continued the research he had begun on East German forests and on this basis, aimed to alert the West German public to the serious danger that industrial emissions like sulfur dioxide posed to forests.[78] His work contributed in part to the *Waldsterbendebatte*, or "forest death debate" in West Germany – a

73 Roth and Murphy, "Competing Factions," 59.
74 Ditfurth *Das waren die Grünen*, 75–76.
75 Helmut Fogt, "The Greens and the New Left: Influences of Left-Extremism on Green Party Organisation and Policies" in *The Greens in West Germany: Organisation and Policy Making*, ed. Eva Kolinsky (New York: Berg, 1989), 89.
76 Hermand, *Grüne Utopien*, 147.
77 Uekötter, *Greenest Nation*, 125.
78 Huff, *Natur und Industrie*, 149.

sudden eruption of environmentalism into mainstream West German consciousness, sparked by a 1981 series in *Der Spiegel*, surrounding the contention that acid rain was destroying German forests.[79] By 1985 over half of West Germans were convinced that by the year 2000 every tree in Germany would be dead.[80] Suddenly, regulations on emissions and pollution began to crank out of Bonn.[81]

Wilhelm Knabe was not just a scientist or a former East German – he became one of the founders of the West German Greens, and it was in large part the "forest death debate" that the Greens made their first political fame, pledging to be the party that would save the German forests. Scores of activists were catapulted into the Bundestag representing the new party on this early wave of electoral success in 1983, including Knabe.[82] Of course, right wing, *völkische* environmentalists, including Herbert Gruhl, were on the front lines of sounding the alarm about the threat posed to the *German* forests, as well. Left and Right could not agree when it came to the fate of people, but when the trees were thought to be in danger, observed one Hans-Jochen Vogel, the head of the SPD at the time, *then* a national movement could be formed.[83] In fact, as Uekötter argues, the forest death debate did more than just about anything to make environmentalism a mainstream political value shared by most parties in the Bundestag,[84] even if the actual level of threat may very well have been grossly exaggerated.[85]

Wilhelm Knabe also played a critical role in bridging the Greens and the beginnings of the grassroots resistance groups in the GDR. As a scientist, he was able to maintain professional ties to the GDR. And he was able to use his status as an elected member of the Bundestag to smuggle a printing press

79 Huff, *Natur und Industrie*, 9, 276.
80 Huff, *Natur und Industrie*, 19.
81 Huff, *Natur und Industrie*, 276.
82 Uekötter, *Greenest Nation*, 114 and Huff, *Natur und Industrie*, 276.
83 Uekötter, *Greenest Nation*, 114.
84 Uekötter, *Greenest Nation*, 114.
85 See Huff's contribution to this volume in which he uses the term "*Waldsterbenhysterie*" instead of *Waldsterbendebatte*. See also Frank Uekötter and Jens Hohensee, eds., *Wird Kassandra heiser? Die Geschichte falscher Ökoalarme* (Stuttgart: Steiner, 2004).

to the members of the environmental library in the Zion Church in Berlin.[86] It was this printing press that ultimately provoked the 1987 Stasi raid which Huff cites as the "beginning of the end" of the GDR. When the Stasi raid happened, one of the first to speak out about it and bring attention to the West – and through Western press, to the East as well – were the leaders of the Green Party in Bonn, including Knabe and Petra Kelly.[87]

Nor was Wilhelm Knabe the sole instance of western Greens actively seeking to aid dissidents in the GDR. His son, Hubertus, was also a member of the Greens and formed the committee that agitated, successfully, for Rudi Bahro to be released from East German prison. In fact, it was through the efforts of both of the Knabes that Bahro found his way to the Green Party as a co-founder and chief ideologue. Knabe was one of the first to identify the link between environmentalism and human rights as having the potential to destabilize Eastern European socialism, and after 1989 published extensively about both the human rights and environmental abuses of the GDR.[88] It is not a coincidence, then, that Hubertus has gone from being a Green Party founder and activist to become perhaps *the* most vociferous advocate for the memory of the victims of state repression in the GDR, even equating them with the victims of the Holocaust, and frequently decrying the lack of seriousness with which the German public takes the crimes of the GDR relative to the crimes of the Third Reich.[89] Nor were the Knabes the only western Greens to engage in real dissidence and resistance work against the East German state: Petra Kelly and other Green leaders such as Otto Schily famously demonstrated in Alexanderplatz in 1983 for peace, an action which earned them big headlines in the West, but which led, ultimately to being banned from the GDR for a time.[90] Other Greens also

86 Huff, *Natur und Industrie*, 357.

87 Huff, *Natur und Industrie*, 358.

88 Hubertus Knabe, *Umweltproteste in Ungarn: Innen- und aussenpolitische Implikationen* (Cologne: Bundesinstitut für Ostwissenschaftliche und Internationale Studien, 1989).

89 See among many sources, Hubertus Knabe, *Die Täter sind unter uns: Über das Schönreden der SED-Diktatur* (Berlin: Propyläen, 2007).

90 Volker Gransow, "A Greening of German-German Relations?" in *The Greens in West Germany*, ed. Eva Kolinsky, 144.

provided crucial contacts for GDR dissidents, including smuggling books and information for them to disseminate from the environmental library,[91] or using quasi-pirate radio stations to broadcast from West Berlin into East Berlin, including Radio 100, also known as *Radio Glasnost*, which was run in part by Siegbert Schefke, who had helped found the environmental library before being expelled from East Germany; *Radio Glasnost* broadcast information about pollution in the GDR as well as helped provide information that helped protest groups coordinate, further strengthening the fusion between environmental activists and the broader dissident movement in the GDR.[92]

Even though, as a parliamentary party, the Greens did not rise above 8 percent and were not part of a ruling coalition until 1998, as a *movement* they exerted great influence over mainstream parties and their politics in the 1980s and 1990s. The CDU, which ruled West Germany from 1982–98, co-opted much of the Green movement, enacting strict environmental legislation itself in the 1980s.[93] This was the legislation that David Blackbourn credits with having saved the environment of the FRG, which was precisely what did not happen in the GDR – the turning point that failed to turn.[94] Undoubtedly, the ideological heterogeneity of the Greens and their clear distance from the Left, including the USSR and East Germany, helped ease the acceptance of green ideas into the mainstream and even the center-right of West German politics, which enabled this turning. Once the more radical "Fundis," led by Ditfurth, were pushed out of the Party's leadership by the more moderate "Realos" led by Fischer,[95] the Greens, or at least their demands for strict environmental regulation, were even more acceptable to the mainstream, including the CDU.[96]

For all these reasons, from the point of view of the East German state and party, environmentalism became weaponized in the German-German

91 Huff, *Natur und Industrie*, 356–66.
92 See Michel Dupuy's contribution to this volume.
93 Uekötter, *Greenest Nation*, 114.
94 Blackbourn, *Conquest*, 340.
95 Mewes, "History of the Green Party," 43–45, and Uekötter, *Greenest Nation*, 139.
96 Uekötter, *Greenest Nation*, 124.

rivalry. As with so many other aspects of this rivalry, the GDR found itself caught playing a game in which the rules were set by the West. For the GDR, the organizing principle since at least 1958, and to some degree 1953, had been to catch up with and overtake the West in terms of per capita consumption.[97] The palace coup that replaced Walter Ulbricht with Erich Honecker in 1971 came about largely because the Party, and the USSR, believed Ulbricht was not succeeding in this task, and saw trouble on the horizon in the unrest in Poland in 1970 if the GDR did not deliver a higher standard of living to its citizens.[98] The West German economic miracle shifted the basis for legitimacy in the German-German rivalry onto the terrain of consumption and domestic living standards, and that had placed tremendous pressure on the SED. Honecker's regime based its legitimacy on the "Unity of Economic and Social Policy," often referred to as "real existing socialism."[99]

The GDR was trying desperately to create a consumer society that could match the material comforts of the Federal Republic, but it was forced to do so within the confines of a socialist planned economy, one lodged within the Soviet international economic system and within the context of the global economic order of the Cold War. One way in which this impacted the East German *environment* was the GDR's lack of access to traditional materials on the world market, due to embargos and a lack of hard currency. Without a cheap supply of wood, aluminum, cotton, etc., it was nearly impossible to mass produce the consumer products that would allow East Germans to enjoy some measure of the "good life" – as their West German neighbors seemed to be enjoying. So, East Germany relied on its synthetics industry – on plastics – to synthesize a socialist consumer culture. East German everyday life

97 Sozialistische Einheitspartei Deutschlands, *Protokoll der Verhandlungen des V. Parteitages der Sozialistischen Einheitspartei Deutschlands* (Berlin: Dietz, 1959), 68.

98 André Steiner, *The Plans that Failed: An Economic History of the GDR*, trans. Ewald Osers (New York: Berghahn, 2010), 143–44.

99 Steiner, *Plans*, 146. Also see Beatrix Bouvier *Die DDR – Ein Sozialstaat? Sozialpolitik in der Ära Honecker* (Bonn: Dietz, 2002); Greg Castillo, *Cold War on the Home Front: The Soft Power of Midcentury Design* (Minneapolis: University of Minnesota Press, 2010).

became defined by plastic – people mostly wore polyester clothes, walked on polyvinyl chloride floors, ate off of polystyrene, played with polyethylene toys, worked on melamine desks, etc.[100]

But, all this plastics production came at a price to the environment: much of the synthetics industry in the GDR dated to the early twentieth century when lignite was used as a feedstock for industrial chemistry. Because many of the chemical plants which produced the plastics, such as Bitterfeld, Leuna, and Buna, were built near deposits of lignite, or brown coal, in an area of Saxony-Anhalt known as the "chemical triangle." Lignite burns much "dirtier" than black coal and especially petroleum when it is processed as fuel and as a petrochemical basis for plastics.[101] In particular, lignite emits high levels of sulfur dioxide, which can cause "acid rain." The reliance of the chemical triangle to create the basis for the East German consumer economy was in part what caused such terrible pollution in the GDR – but this situation came about because 1) earlier capitalist modes of production had bequeathed to the GDR a highly polluting infrastructure and 2) the global cold war competition left the East Germans with little choice but to rely on plastics to keep up with their more affluent neighbors and competitors to the West.

In fact, the GDR did not deny or ignore the fact that there was industrial pollution within its borders: its official Party line was that the pollution its factories spewed were a leftover from the capitalist era, and in particular, *capitalism's* rapacious exploitation of nature for profit,[102] and that socialism was trying to *clean up* the mess capitalism had made, while also building a comfortable society for its residents. Once the old machinery from the capitalist era had been replaced by socialist means of production, which would be made in harmony with nature, pollution would be a thing of the past, root and branch, according to SED ideologues such as Kurt Hager and Alfred Kosing.[103]

100 See Eli Rubin, *Synthetic Socialism: Plastics and Dictatorship in the German Democratic Republic* (Chapel Hill: University of North Carolina Press, 2008).
101 See Chaney, "Chemical Landscape Transformed," 145–146.
102 See Huff, *Natur und Industrie*, 49–50 and 191.
103 Huff, *Natur und Industrie*, 191.

Indeed, East Germany was concerned with industrial pollution and its remediation well before West Germany, and most western countries, as Tobias Huff notes in his contribution to this volume. The GDR had enshrined environmental protection in its 1968 constitution, in Article 15.[104] This constitution then became the basis for numerous laws and legal structures, such as the formation of a Ministry of Environmental Protection and Water Management in 1971 (West Germany did not create an Environmental Ministry until 1986),[105] which set stringent limits on emissions and imposed fines on those factories that violated the limits. According to Michel Dupuy[106] and Tobias Huff, the GDR had a track record of being very open about discussions concerning pollution in the 1960s and 1970s, and often sponsored "forest clean up" actions, for example, through the SED Youth Organization, the FDJ (*Freie Deutsche Jugend* [Free German Youth]).[107]

The problem for the GDR, and the reason that its industry continued to pollute, was not a result of negligence or carelessness, nor was it anything to do with inherent predispositions to "growth," prometheanism, or "gigantomania" in socialism. It was instead a twofold problem: first, it was never able to switch from the highly polluting lignite to the much cleaner petroleum as a feedstock for its consumer economy; and second, it was unable to enforce the environmental regulations that it did pass with as much teeth as western countries were able to.

Why was the GDR not able to switch to cleaner burning petroleum? Officials had planned on either retrofitting lignite-based factories to use petroleum, or in some cases closing older factories, such as the notorious polluting Espenhain facility, and building new petroleum based production sites,[108] such as the VEB Film Factory in Wolfen.[109] But because the

104 Nationalrat der DDR, *Verfassung der DDR* (Berlin: Staatsverlag der DDR, 1968); Article 15; also Huff, *Natur und Industrie*, 167.

105 Blackbourn, *Conquest of Nature*, 338–39 and Huff *Natur und Industrie*, 183–84.

106 See Dupuy's contribution in this volume.

107 Huff, *Natur und Industrie*, 287.

108 Huff, *Natur und Industrie*, 194.

109 Huff, *Natur und Industrie*, 195.

GDR could not afford to buy oil on the world market using hard currency, it relied on petroleum from the Soviet Union, which was be delivered through the "Friendship Pipeline" built through Eastern Europe with its terminus at the refinery in Schwedt, on the GDR-Poland border. Initial estimates were that the GDR would receive 40–50 million tons of oil a year, enough to leave lignite behind entirely.[110] Turned on in 1963, the pipeline reached a maximum capacity of only 19 million tons a year, not enough to transform the GDR's energy and industrial economy to being completely lignite-free.[111]

And in 1980, this amount dropped from 19 to 17 million tons a year of petroleum, as a result of international oil shocks and East German-Soviet tensions over East German use of the oil.[112] The leadership of the GDR knew right away how devastating this cutback would be; when informed of the decision, Honecker declared to the Soviets that this change threatened "the very existence of the GDR."[113] He was right, as Raymond Stokes has argued,[114] it had a devastating effect on both the GDR's economy *and* ecology. Factories that were supposed to be transitioned away from lignite had to stay on lignite, such as Espenhain, at the very moment that industrial production was ramping up to churn out consumer goods, many of which that were made of plastic. As a result, lignite usage skyrocketed from 250,000 tons a year to 310,000 tons a year, an increase of 25 percent. This meant a visible, tangible increase in acid rain, smog, dust, and the pollution of groundwater, erosion, and even the removal of villages and towns through the expansion of lignite

110 Rubin, *Synthetic Socialism*, 29; Rainer Karlsch and Raymond Stokes, *The Chemistry Must Be Right: The Privatization of Buna Sow Leuna Olefinverbund GmBH, 1990–2000* (Berlin: Buna Sow Leuna Olefinverbund GmBH, 2001), 32.

111 Also see Stokes, "From *Schadenfreude* to Going-out-of-Business Sale: East Germany and the Oil Crises of the 1970s," in *The East German Economy, 1945–2010: Falling Behind or Catching Up?*, eds. Uta Balbier and Hartmut Berghoff (New York: Cambridge University Press, 2013), 131–145.

112 Stokes, "Oil Crises," 137.

113 Rubin, *Synthetic Socialism*, 29; Karlsch and Stokes, *Chemistry*, 40; and Stokes, "Oil Crises," 139.

114 Stokes, "Oil Crises," 142.

mining now necessary to feed the increased demand for brown coal in the GDR.[115]

At the same time that pollution was climbing in the late 1970s and early 1980s, the Environmental Protection Ministry found that it was largely powerless to punish companies that violated emissions limits, because the fines it levied on polluting factories were minute in comparison to the costs which those factories would have had to have incurred to purchase and install new filters or smokestacks.[116] For example, according to Huff, the cost of installing a sulfur dioxide filter at Leuna was estimated at 2 billion marks, and would likely have only reduced emissions by 2 percent.[117]

From the perspective of the GDR, then, the problem was not one of "growth" per se, nor was it in any way a problem with socialism or Marxism itself – the problem was the capitalism had left a mess to clean up that was proving much more expensive than anyone anticipated. Seen in this light, the newly trendy attacks on growth coming from the West were rich, indeed, and any mirroring of such a critique of growth within socialism simply missed the bigger point: capitalism was throwing up every conceivable roadblock to socialism, and trying to convince socialist East Germans that they should dismantle the industrial economy they struggled so hard to build up was just another, particularly cunning, move to distract people from what the ultimate goal of communism had to be – the satisfaction of people's wants and needs in an equitable way. Harry Nick, a top Party official in charge of economics and technology, stated plainly that "Communism without growth would not be some other form of communism, but not communism at all" and claimed that the *Limits to Growth* was the product of a "bourgeois ideology" to use environmentalism to "replace the socialist revolution."[118] Others shot back at the attempt to criticize growth, including the philosopher Alfred Kosing, who stated in his 1988 book *Sozialismus und Umwelt* [Socialism and the Environment] that no matter what, "dynamic economic growth is

115 Huff, *Natur und Industrie*, 193.
116 Huff, *Natur und Industrie*, 198–99. Also Chaney, "A Chemical Landscape," 147.
117 Huff, *Natur und Industrie*, 198.
118 Huff, *Natur und Industrie*, 201.

the basis for the progress of a socialist society."[119] Any thought of scaling *back* their program of consumer goods production was totally anathema for East German leaders – at so many junctures, from the 1953 uprising to the near uprising that took place during the 1977 coffee shortage, the East German citizenry made it overwhelmingly clear what they wanted from the SED, and that was a modern, comfortable, consumer lifestyle at least somewhat on par with the West.

From the point of view of the GDR, the sudden and somewhat late-to-the-game concern about the environment in the West smacked of a kind of cultural snobbery: just as soon as East Germans finally started to accumulate nice stuff, West Germans declared "greenness" to be the measure of the good life instead of the acquisition of consumer goods. After decades of touting their success at being a materialistic society, and using their economic "growth" as a way of shaming the lagging socialist economy (for not having enough growth), the West Germans were now shaming the East Germans for being *too* obsessed with growth, consumption, and materialism, instead of the new vogue for anti-materialism and environmentalism. East Germans smelled hypocrisy, and even claimed to see through the forest death debate as manufactured hysteria: in fact, Hans Reichelt decried the *Waldsterbendebatte* as precisely that – "hysteria" – as well as a "manipulation of the masses through the press"[120] and Erich Honecker himself declared in an interview with *die Zeit* "there are no forests dying here."[121]

Even with their suspicions of western environmentalism, the SED was open, at first, to working with the Greens as potential allies: in 1983 Honecker himself reached out to Kelly, Schily, and other Green leaders with an offer of a state visit to East Berlin, which in fact took place, leading to a promise of a German-German symposium on the environment, and another on peace, even after the stunt they pulled in Alexanderplatz.[122] But this did not last, as East German leaders realized that the Green movement contained strong anti-Soviet and anti-Marxist elements, that its organizing principle of limiting

119 Hermand, *Grüne Utopien*, 147.
120 Huff, *Natur und Industrie*, 19.
121 Huff, *Natur und Industrie*, 7.
122 Gransow, "Greening," 144–46.

growth posed an existential threat to the GDR, and that the Green movement itself was very much an exit ramp for former Marxists who had come to care more about trees than about people. The GDR leadership came to see that the FRG would use environmentalism as a cudgel with which to gain the upper hand in the battle for moral superiority between West and East.[123]

Thus, in 1982, the GDR Ministers' Council ended the open discussion of environmental pollution, declaring all information about the state of the environment in the GDR to be a state secret.[124] This, like so many of the repressive actions of the GDR state, ended up being a self-fulfilling prophecy. Repressing information about environmental pollution led directly to the growth of *Basisgruppen* – grassroots, underground resistance groups – which in turn strengthened the assumption in the West that environmentalism and liberal democracy went hand in hand, and that the pollution of one led to the repression of the other, or vice-versa. In the minds of West German Greens, ex-68ers and many, many other West Germans from across the ideological spectrum, "greenness" became the defining characteristic which cemented the West's superiority over the East, which gave them the "environmental moral high ground vis-à-vis the GDR" (*umweltmoralische Lufthohheit gegenüber der DDR*), according to Huff.[125] By the end of the 1980s, the West German public had already internalized a kind of green condescension towards the East, and sensational reports of unimaginable environmental pollution in the other Germany were very popular among West German audiences, culminating in the famous documentary "Bitter News from Bitterfeld," containing footage shot surreptitiously by underground environmental groups showing damage to forests near the Bitterfeld factory, which was aired on West German TV in 1988.[126]

123 Uekötter makes this point in "Environmentalism, Eastern European Style: Some Exploratory Remarks" in *Umweltgeschichte(n): Ostmitteleuropa von der Industrialisierung bis zum Postsozialismus: Vorträge der Tagung des Collegium Carolinium in Bad Wiessee vom 4. Bis 7. November 2010*, eds. Horst Förster, Julia Herzberg and Martin Zückert (Göttingen: Vandenhoeck & Ruprecht, 2013) 241–254, 248–49.

124 Uekötter, *Greenest Nation*, 135.

125 Huff, *Natur und Industrie*, 19.

126 Uekötter, *Greenest Nation*, 136.

The Greens and their allies were hopeful that the meaning of the 1989 *Wende* was that Soviet state socialism collapsed because of opposition to its exploitation of nature and its embrace of growth and consumption, seeing in the grassroots groups that sparked the revolution – including some they helped to foster – validation of their mission and ideology. They were sorely disappointed. In the first free elections in the GDR, in the spring of 1990, the Greens fared terribly. In the first federal elections of a reunified Germany, in December of 1990, the Greens failed to meet the 5 percent threshold for the first time in almost a decade, dragged down by vanishingly few votes in the "new federal states" of the former GDR.[127] Many western Greens, especially those most influenced by Bahro and thus opposed to "growth" expressed bitterness towards the East Germans who voted not for less growth and consumption, not for a third way between the twin Molochs of capitalism and communism, but squarely for growth, consumption, and capitalism.

Thus, the elections of 1990 revealed this fact: that East Germans *were* in fact concerned mainly with a materialistic, consumer-driven lifestyle, which was dependent on growth, just as the SED had argued for decades. They just picked a more successful model of delivering consumerism and growth, which they thought would deliver consumption and growth to them, namely capitalist West Germany led by the CDU. To this day, the electoral base of the Greens is in the West, not the former GDR. The irony of this, of course, is that one of the great accomplishments of reunified Germany was the massive environmental remediation of the former GDR, a result of the mainstreaming of Green politics in the pre-Wende FRG, and at tremendous expense to (mostly western) taxpayers.

***

The phenomenon of "greenwashing" has become more and more the subject of commentary in recent years.[128] It implies the tendency of companies or

127 Uekötter *Greenest Nation*, 139.

128 See Hartmut Berghoff and Adam Rome, eds., *Green Capitalism? Business and the Environment in the Twentieth Century* (Philadelphia: University of Pennsylvania Press, 2017), especially Rome, "The Ecology of Commerce: Environmental History and the Challenge of Building a Sustainable Economy"; also Fred Magdoff and John

organizations to adopt "green" policies mainly for purposes of public relations, while continuing to pollute, especially in places where mainstream culture will not see – for example, in the developing world. This critique implies that the West has been able to transcend the antagonism between a consumption-driven economy based on endless growth, on the one hand, and the environment on the other, largely by displacing much of the industrial production, and pollution, to the developing world, with its lax environmental (and labor) protections. Obvious forms of pollution – smog, trash dumps, discolored or noxious tap water, oil spills, etc. – have been largely banished from the immediate, daily lives of Westerners, reducing the tinder that might light another radical environmental movement. Or more accurately, it has been banished from the lives of privileged, often white, middle-class Westerners – the example of Flint, Michigan, being emblematic for the kind of environmental and racial/socioeconomic double standard that adheres in many "first-world" nations.

For example, the West has for a long time sold its trash to the Third World; in fact, West Germany sold *its* trash to East Germany.[129] Indeed, much of the capitalist world's trash simply ends up in the middle of the ocean, where no one has to see it.[130] Out of sight, out of the political mind.

Bellamy Foster, *What Every Environmentalist Needs to Know About Capitalism: A Citizen's Guide to Capitalism and the Environment* (New York: Monthly Review Press, 2011); Naomi Klein, *This Changes Everything: Capitalism vs. the Climate* (New York: Simon and Schuster, 2014).

129 See Christof Mauch, ed., *Out of Sight, Out of Mind: The Politics and Culture of Waste*, RCC (Rachel Carson Center) Perspectives: Transformations in Environment and Society 2016/1 (Munich: Rachel Carson Center, 2016); especially Simone Müller "The 'Flying Dutchmen': Ships' Tales of Toxic Waste in a Globalized World" and Richard Grant "The 'Urban Mine' in Accra, Ghana"; also Stephanie Foote and Elizabeth Mazzolini, eds., *Histories of the Dustheap: Waste, Material Culture, Social Justice* (Cambridge, MA: MIT Press, 2012). On East Germany and West German trash, see Stokes "Oil Crises," 143; he is citing a Ministers' Council document on a 20-year contract with West Berlin to let the FRG dump its trash in East Germany.

130 Kiara Alfonseca, "The Great Pacific Garbage Patch, full of plastic, keeps growing," accessed May 11, 2018, <https://www.nbcnews.com/news/us-news/great-pacific-garbage-patch-full-ocean-plastic-keeps-growing-n859276>.

It was this displacement that East Germany could not do – whatever "greenwashing" the SED attempted to implement was, like so much of its propaganda, clumsy and easily contradicted by the immediate experience of citizens who could see, smell and taste the chemical legacy of the industrialization that supported their consumer lifestyle. Just like the fact that capital can move across borders and oceans but people often cannot, so too were capitalist nations able to move pollution across oceans whereas socialist countries could not move their heavy industry or their trash to other nations, and it may be that this feat was what won the Cold War for the capitalist nations – or at least proved to be their decisive advantage. This is the bigger purpose in kicking the tires of a "Green legend" such as has been constructed over the last several decades – it conceals the greenwashing that allows pollution to continue, out of sight and out of mind.

Hovering over these issues is another the question that must be considered: Had the GDR built a socialist economy and society under more auspicious circumstances, with more capacity and capital to implement green technology, would it actually have created a world in which the *Wechselwirkung* between humans and the environment, which Marx foresaw, would have come to pass? In a way, this hypothetical question is an even bigger one – would socialism have succeeded if not weighed down by the albatrosses of the postwar conditions, Soviet occupation and Stalinist policies in general, all of which sapped vital resources that were needed to do things like install cleaner burning industrial technology?

In another way, it is part of an even bigger question than that – is the problem that capitalism is *a priori* to growth, or is growth *a priori* to both capitalism and socialism? Had socialism triumphed over the globe instead of capitalism, would it not have also led to global warming? Would it not have also produced a massive trash island in the middle of the ocean? Is socialism not guilty of greenwashing only because it was not able to greenwash, but would have if it could have? These are very, very big questions, which cannot be answered here – but they were central to the ecology of ideologies, human and non-human actors, and states during the Cold War, they shaped the course of both the Cold War and the environment itself, and they need to be re-asked now with the perspective of time's passage.

The point here is not an apologia for the GDR – its pollution was very real. It is to reexamine the narrative constructed out of the Cold War which has tied pollution to "prometheanism" and both of these to socialism inherently. This was an artifice of the Cold War, the complicated ideological and geopolitical "ecosystem" that formed amongst the West German Left, elements of the Right, and the GDR, and one which formed a "Green Legend." This narrative needs to be re-examined in more detail; this essay merely seeks an opening to such a larger project.

JULIA E. AULT

Aquatic Conundrums: The GDR's Water Woes and Soviet Bloc Cooperation, 1963–1989

Introduction

In 1978, members of the Council for Mutual Economic Assistance (Comecon) met in Moscow to announce the establishment of the *Interwodootschistka*, or "Inter-Water Treatment." The organization institutionalized "scientific-technological basis to improve production capacities" and aimed to "protect the needs of these countries through a quicker and more effective solution to the problem of environmental protection."[1] The GDR (German Democratic Republic, East Germany) touted the *Interwodootschistka*'s potential. Experts in the GDR admitted to difficulties in distributing the necessary water to its major industrial centers and lauded the "systematic cooperation in the development and production of quality facilities" that the new organization proposed.[2] The *Interwodootschistka* promised answers to domestic water shortages and pollution while placing the GDR at the forefront of a conversation about technological innovation within the Soviet bloc. Water practically tied together socialist countries through similar economic and industrial structures, literally linked them through shared waterways, and symbolically created an ecology or interconnected set of countries and problems in socialism.

1 Bundesarchiv (BArch) DK 5/2248, "Abkommen über die Gründung der Internationalen Wirtschaftsgemeinschaft 'Interwodootschistka,'" Moscow, 1978.

2 BArch DK 5/1046, "Rede zum Tagesordnungspunkt: 'Interwasserreinigung,'" April 23, 1975.

East German leaders recognized the devastating effects of inefficient water use as early as the 1950s and sought to address them in a comprehensive way in the following decade. With the Socialist Unity Party's (SED) approval, the GDR formulated a multipronged response to its water conundrums. Officials first implemented a series of new economic policies in the 1960s that incentivized pollution abatement. They also looked to update sewage, purification, and water recycling for industrial and domestic needs and further enshrined environmental protection into the constitution in 1968.[3] Movements in western countries also influenced the GDR and the Soviet bloc, but they did not simply mimic the West in their response to water management and the environment. Instead, they drew on Marx's writings: "Nature always forms the basis of human existence, but what attitude people have towards nature ... depends above all on the given economic social formation."[4] East German and Soviet bloc officials recognized the tensions between nature and industry, but developed strategies internally as well as in the international community to improve conditions, broadly, and water quality and quantity, specifically.

Still, East Germans faced a conundrum shared across the Soviet bloc, namely, raising productivity while also engaging in environmental discourses and protection. As East Germans advocated for enhanced cooperation through organizations such as the *Interwodootschistka*, they also established the GDR's position as a scientific and technological pioneer in water management. While some work has already been done on German-German water relations, the GDR's position in the Soviet bloc invites further inquiry into relations between socialist countries on the intertwined matters of water and the environment.[5] Water not only defined East

3 Mary Fulbrook, *The People's State: East German Society from Hitler to Honecker* (New Haven, CT: Yale University Press, 2005), 37.

4 Karl Marx, *Grundlagen des historischen Materialismus* (Berlin: Dietz, 1977), 91, cited in Horst Paucke and Adolf Bauer, *Umweltprobleme: Herausforderung der Menschheit* (Berlin: Dietz, 1979), 5.

5 Astrid M. Eckert, "Geteilt aber nicht unverbunden: Grenzgewässer als deutsch-deutsches Umweltproblem," *Vierteljahreshefte für Zeitgeschichte* 21, no. 1 (2014): 69–100; Tim Grady, "A Shared Environment: German-German Relations along the Border, 1945–72," *Journal of Contemporary History* 50, no. 3 (2015): 660–679.

German-Soviet relations, but also partnerships with its eastern European neighbors – especially Poland and Czechoslovakia – through bilateral and trilateral agreements. Ongoing negotiations between the three about water rights and quality framed many of the discussions in a region in which all three polluted one another and yet were obligated to the same economic and diplomatic system.

Many scholars have written about the environment in the GDR and other Soviet bloc countries, but questions remain about the connections and challenges.[6] Tobias Huff's recent work provides the most comprehensive treatment of the GDR's environmental policies and protest, but does not systematically explore its relationships other countries.[7] Other scholars have devoted attention to the Soviet Union, particularly Douglas Weiner, and eastern bloc countries, such as Barbara Hicks' work on Poland, but these works do not address the relationship between nations. Newer edited volumes have delved into the rise of international environmental movements in the 1970s and offered case studies of various countries, but have largely ignored the relationship between Soviet bloc countries.[8] Examining the interactions between Comecon nations over the topic of

6 See Tobias Huff, *Natur und Industrie im Sozialismus: Eine Umweltgeschichte der DDR* (Göttingen: Vandenhoeck & Ruprecht, 2015); Barbara Hicks, *Environmental Politics in Poland: A Social Movement between Regime and Opposition* (New York: Columbia University Press, 1996); Eagle Glassheim, *Cleansing the Czechoslovak Borderlands: Migration, Environment, and Health in the Former Sudetenland* (Pittsburgh, PA: University of Pittsburgh Press, 2016); Edward Snajder, *Nature Protests: The End of Ecology in Slovakia* (Seattle: University of Washington Press, 2008); Barbara Jancar-Webster's *Environmental Action in Eastern Europe: Responses to Crisis* (Armonk, NY: M. E. Sharpe, 1993); Douglas R. Weiner, *A Little Corner of Freedom: Russian Nature Protection from Stalin to Gorbachev* (Berkeley: University of California Press, 1999); and Douglas Weiner, *Models of Nature: Ecology, Conservation and Cultural Revolution in Soviet Russia* (Pittsburgh, PA: University of Pittsburgh Press, 2000).

7 Huff does address West Germany's influence on East German affairs, and discusses the evolution of a treaty with Czechoslovakia, but the framing of the project around *Waldsterben* [forest dieback] leads him to focus more on air pollution than water.

8 See Christof Mauch, Nathan Stoltzfus, and Douglas R. Weiner, eds., *Shades of Green: Environmental Activism around the Globe* (Lanham, MD: Rowman & Littlefield

environmentalism reveals that these nations jockeyed for technology and prestige while also dealing with economic and environmental needs, all while supposedly in cooperation with one another. They formed a kind of "ecology" of socialism. Water, with its tendency to both delineate and to cross national boundaries, is one way of better understanding this ecology.

East Germans' relationship to water – as commodity, component of production, and source of workers' well-being – influenced domestic policy as well as the GDR's position within the Soviet bloc. Increased industrial demands, along with the SED's self-imposed pressure to provide for its citizens, led to a critical water situation in the 1960s and 1970s that the SED felt it had to address. In turn, the SED's concerns about water management shaped the GDR's relationship with other socialist countries, as it championed environmental protection efforts both domestically and internationally and promoted the GDR as a water expert in the bloc. Despite these aspirations, chronically high pollution levels and water shortages heightened tensions between eastern bloc neighbors, forcing a transition away from bloc-wide cooperation and toward more limited agreements. In the 1980s, water – and the environment more broadly – became a matter of national security for the *Stasi* (Ministry for State Security, MfS). Ultimately, water tied together Soviet bloc countries economically and environmentally, which compelled members to adopt shifting strategies that reflected ongoing problems in a changing political environment.

Water Woes

The GDR's lack of substantial waterways troubled East German politicians as early as the 1950s. Its high population density (157 residents per square kilometer) and "highly developed industry and agriculture" placed

Publishers, 2006); J. R. McNeill and Corinna R. Unger, eds., *Environmental Histories of the Cold War* (New York: Cambridge University Press, 2010).

considerable demands on the water supply, as "one of the most strained of all industrialized countries."[9] Water was critical to the economy, agriculture, and industry, as well as moving goods and expanding production. Water, however, was also essential for provisioning the population and providing East Germans with the quality of life for workers in a workers' state, in the home and for recreation. Fishing, swimming, and other pastimes required clean water. Water also figured prominently into German conceptions of nature and was crucial to the SED's project of building nationhood and loyalty in the GDR.[10] By 1969, when 66 percent of East German waterways were categorized as "highly" or "inadmissibly" polluted (the lowest two categories out of four), the state expanded water conservation and protection initiatives.[11]

The economy obviously drove many policies, but water's place in "socialist *Landeskultur*" [roughly translated as "landscape care"] and "socialist environmentalism" also shaped the GDR's approach. In the 1960s, officials coined the term socialist *Landeskultur*, blending older notions of *Heimat* and conservation into evolving ones about environmentalism. *Landeskultur* sought to improve the land and nature while inculcating sense of East German nationhood and economic prosperity.[12] In the 1970s, socialist environmentalism generally overtook *Landeskultur* in official parlance, though texts continued to use both and rarely distinguished between them. Nevertheless, socialist environmentalism focused less on Heimat and nation and instead emphasized "the wellbeing of the workers" and the improvement and effective use of the ground, water, air, plants, and animals through scientific and

9 BArch DK 5/1738, "Die langfristige Entwicklung von Hauptrichtungen des Umweltschutzes in der DDR," August 1974.

10 See Jan Palmowski, *Inventing a Socialist Nation: Heimat and the Politics of Everyday Life in the GDR, 1945–1990* (New York: Cambridge University Press, 2009); Scott Moranda, *The People's Own Landscape: Nature, Tourism, and Dictatorship in East Germany* (Ann Arbor: University of Michigan Press, 2014), 2.

11 BArch DC 20/I/3/0715, Prognosegruppe Abprodukte und sozialistische Landeskultur, "Prognose: Industrielle Abprodukte und planmäßige Gestaltung einer sozialistischen Landeskultur in der DDR," September 1968, 19.

12 Moranda, *The People's Own Landscape*, 63.

technological advancement.[13] The goal was to "bring into agreement workers' material and cultural endeavors and the requirements of the environment."[14] In both conceptions, water was of great importance, and how to deal with it led to conflicts between different interests within the party and state.

The protection of and engagement with nature was also a longstanding component of German identity, which added cultural significance to maintaining clean waterways. Connections between nationalism and nature dated back to the nineteenth century, when preservationists "imagined Germany in the natural landscape" while also reacting to the changes that industrialization and urbanization wrought.[15] Germanness was tied to the physical landscape, which in turn became a justification for preserving it. The SED continued this trend, attempting to instill a particular sense of East German identity through nature.[16] The SED's National Front and Cultural League established campaigns and competitions for East Germans to engage with it, such as maintaining local preserves and hiking trails. These efforts included "maintaining clean waters" for recreation, such as swimming, fishing, and boating, and for the sake of biodiversity. Nature was incredibly popular, with some 60,000 members in the Cultural League's "Friends of Nature and Heimat" organization, and the popularity of such organizations formed an impetus for the GDR to worry about its water.[17]

Early policies, however, focused primarily on basic provisioning of the population and for the functioning of the economy. Its emphasis on industries that required large amounts of water – specifically heavy industry and

13 BArch DK 5/1831, "Information über den Stand die Enwicklung des Umweltschutzes in der DDR," April 5, 1972.

14 Manfred Fiedler, "Vorbemerkung," in *Mensch und Umwelt: Aus dem Protokoll eines Kolloquiums des Kulturbundes der DDR* (East Berlin: Kulturbund der DDR, 1975), 3.

15 Thomas M. Lekan, *Imagining the Nation in Nature: Landscape Preservation and German Identity, 1885–1945* (Cambridge, MA: Harvard University Press, 2004), 5.

16 See Palmowski, *Inventing a Socialist Nation*.

17 BArch Stiftung Archiv Parteien und Massenorganisationen (SAPMO) DY 27/5649, "Vorschlag: Gründung Gesellschaft für Heimat und Umwelt im Kulturbund der DDR," August 21, 1974, Büro Haines.

the production of chemical products – further strained the GDR's relative water shortage. Additionally, water cooled power plants and other industrial processes and was essential for agriculture. Often, industries exceeded their water allotments.[18] Moreover, runoff and the dumping of byproducts generated pollution, which required purifying water before it could be reused or drunk.[19] Addressing this situation meant building reservoirs and channels, expanding infrastructure, and refining mechanical and chemical treatments.[20] As early as 1952, the Council of Ministers invested nearly 3 million DM in water management to alleviate chronic shortages in Berlin. Twelve additional projects totaled roughly 2.3 million DM.[21] Plans included building reservoirs for drinking water in the Harz and Erzgebirge, treating agricultural runoff, draining the Oder River plain, managing the water table in regions with open-pit coal mining, and planning for flood control.

Demand on the water supply necessitated the new "Water Law" in 1963, which became a foundational piece of legislation, and remained in effect for nearly 20 years.[22] Officially titled "The Law on the Protection, Use, and Maintenance of Waterways and the Protection against Flooding Dangers – Water Law – from April 17, 1963," it specified immediate areas of concern and plans to address them. It centralized water management, prioritized supplying the population, required industries use water more efficiently, and limited pollution from runoff. It explicitly encompassed the protection of running and standing water, ground water, and coastal regions, including beaches.[23] Lawmakers particularly noted the need to

18 BArch DY 30/J/IV 2/2/1944/01, "Grundsätze für das neue Wasstergesetz," May 4, 1982, 2–3.

19 BArch DC 20/19102, Dr. Werner Titel, "Die Entwicklung der soz. Landeskultur in der DDR – Ausdruck der kontinuierlichen Politik der SED zur Gestaltung der entwickelten soz. Gesellschaft," January 20, 1971.

20 BArch DK 4/427 Johannes Rochlitzer, Amt für Wasserwirtschaft, "2.1 Reinhaltung der Gewässer."

21 BArch DC 20/I/3/0143, Ministerrat der DDR, "Beschluss vom 3. Oktober 1952."

22 A new Water Law was passed in 1982 and remained in effect until 1989.

23 BArch DK 5/540, "Gesamtüberblick über die Vereinigung der Gewässer durch Mineralöle und deren Nebenprodukte und Massnahmen zur Verhütung derartiger Verunreinigungen in der DDR," undated.

protect water from crude oil spills and runoff from refineries. The Council of Ministers therefore paired the Water Law with a new Office for Water Management in 1963, standardized protections against such crude oil spills in 1965, and then made them binding in 1969.[24]

The New Economic System (NES), also introduced in 1963, and then the Economic System of Socialism (ESS), mirrored changes in thinking about the importance of water to the GDR.[25] While centralization of water management did not neatly map onto the NES's emphasis on permitting various branches of industry more autonomy, it did align with the NES's introducing "economic levers" to increase efficiency. It introduced prices, costs, wages, interest, profits, and other "market mechanisms" to reform the economy, and because *Abwassergeld* was essentially a fine, it fit into these new methods. Yet the NES did not succeed in kickstarting the economy, and so by the late 1960s, some planners and SED officials supported moving away from it, turning toward long-term forecasts to determine production and consumption. With this shift to the ESS, technology, science and engineering remained priorities, and an integral part of water protection.

Water's place in the SED's conception was not purely economic, but also a matter of providing East Germans a certain quality of life that included natural beauty and health as well as material wellbeing. In the 1968 "Prognosis Report on Industrial Byproducts and the Planned Construction of a Socialist *Landeskultur* in the GDR," author and later the GDR's first environmental minister, Werner Titel, argued for protection from multiple angles. Titel demonstrated myriad ways that water shortages and pollution posed challenges to the SED project, economically and socio-politically. In the report, he claimed, "The economic losses through damages, additional expenses, and failure to reuse materials from byproducts in the GDR costs on the order of two billion Marks annually." Additionally, that sum did "not include the immeasurable harm to the quality of human life in

24 BArch DK 5/540, "Gesamtüberblick."

25 André Steiner, *The Plans that Failed: An Economic History of the GDR* (New York: Berghahn Books, 2010), 110.

terms of hygiene and public health as well as the defacement of nature."[26] In calling for the regulation of water, within the concept of *Landeskultur*, Titel's report laid out practical and the ideological foundations for water protection in a socialist state.

Titel placed the GDR in the larger context of the Soviet Union and the socialist bloc, implying that it would not want to be left behind. In particular, he cited Brezhnev's speech to the Central Committee of the CPSU and the Supreme Soviet of the USSR on the fiftieth anniversary of the October Revolution, in which Brezhnev emphasized "how important it is to protect nature and nurture and increase its riches." Brezhnev pushed traditional conservationism, however, saying "The economic and prudent use of natural riches, care for the soil, forests, and rivers as well as the protection of the air and plant and animal worlds – all that is our intrinsic communist task."[27] In his conception, nature was for the building of communism – serving workers, bettering their lives, and strengthening the economy – rather than a balance between the nature and humans. The Soviet Union's dubious environmental record reflected this utilitarian approach, as seen in the inadvertent draining of the Aral Sea, a result of diverting rivers to irrigate cotton fields.[28]

Beyond the Soviet bloc, Titel witnessed the international uptick in environmental regulation and hoped to situate the GDR in that trend. He especially incorporated Swedish accomplishments, because while it was not a socialist country, it was on friendlier terms with the Soviet bloc than other western European countries. It was also an international leader in environmental protection, having led an admirable "public enlightenment campaign to fight the pollution of air and water." In 1968, the Swedish

26 BArch DC 20/I/3/0715, Prognosegruppe Abprodukte und sozialistische Landeskultur, "Prognose: Industrielle Abprodukte und planmäßige Gestaltung einer sozialistischen Landeskultur in der DDR," September 1968.

27 Cited in BArch DC 20/I/3/0715, Prognosegruppe Abprodukte und sozialistische Landeskultur, "Prognose: Industrielle Abprodukte und planmäßige Gestaltung einer sozialistischen Landeskultur in der DDR," September 1968, 7.

28 Douglas R. Weiner, *A Little Corner of Freedom: Russian Nature Protection from Stalin to Gorbachev* (Berkeley: University of California Press, 1999), 359, 415.

government had already introduced a 20-point plan costing the equivalent of 13 billion (East German) Marks to restore environmental conditions by 1975. Perhaps with a touch of envy, Titel noted, "With this program, Sweden will be a leader in Europe, if not the world."[29] Titel, therefore, urged the Council of Ministers to take up environmental protection, so that the GDR would be perceived as a leader in the field both inside and outside the Soviet bloc and solidify an international recognition is so badly sought.[30]

Drawing on the international moment and German trends of conservation, the evolution of socialist *Landeskultur* and socialist environmentalism crucially shaped the SED's understanding of water. Titel conceived of *Landeskultur* as a "system of societal measures for the planned maintenance and improvement of the foundations of natural living conditions and production – soil, water, air, plant and animal life – in the citizens' interest and the continuing development of the economy as an optimal design for the natural habitat."[31] Nevertheless, the report's basis lay in "lines of development in the economic structural complexes" and production. In other words, the primary motivator – and the way to sell environmental protection to the Council of Ministers – concentrated on the "further increase of industrial and agricultural production."[32] While Titel and a handful of others touted socialist *Landeskultur* or environmentalism that included a range of reasons for action, their strongest arguments to the ministers were framed in economic terms.

A supplement to the prognosis report, along with the *Landeskultur* Law of 1970, specified water management priorities in the near future, most especially pollution and shortages. In order to both accommodate economic growth and alleviate poor water quality, the revised report emphasized

29 Weiner, *A Little Corner of Freedom*, 7.

30 William Glenn Gray, *Germany's Cold War: The Global Campaign to Isolate East Germany, 1949–1969* (Chapel Hill: The University of North Carolina Press, 2003), 147–153.

31 BArch 20/I/3/0715, Prognosegruppe Abprodukte und sozialistische Landeskultur, "Prognose: Industrielle Abprodukte und planmäßige Gestaltung einer sozialistischen Landeskultur in der DDR," September 1968, 8.

32 BArch 20/I/3/0715, "Prognose: Industrielle Abprodukte," 8.

essential universal new technologies and more efficient use of water in industrial and agricultural processes.[33] It also introduced new measures for industries that polluted, drawing on the ESS's methods. In 1969 and 1970, a pilot project was run in the district of Halle, in which economic incentives were applied to water cleanup. In fact, if factories polluted above permissible levels, they were "sanctioned" with a "not insignificant" fine called *Abwassergeld* as an "economic stimulus."[34] The Office of Water Management then put incentives (or disincentives) in place to regulate the "enterprises' relationship with the natural environment," meaning essentially fines for surpassing legal maximums.[35]

Strictures on East German industries further highlighted water's importance to the economy, the well-being of the state and its people. For many industries, however, paying a fine was more convenient than dealing with the pollution, or it was not enforced.[36] The results were clear. In Halle, dangerous levels – as much as 72.9 percent of waterways were "highly polluted" in 1980 – remained for decades.[37] In the wake of these laws and the new 1968 constitution that guaranteed the right to a clean environment, the *Landeskultur* Law of 1970 incorporated the "use and protection of waters" into a conception of socialist environmentalism that viewed science and technology as the solution to the GDR's limited

33 BArch DC 20/I/3/0715, "Nachtrag zur Prognose: Industrielle Abprodukte und planmäßige Gestaltung einer sozialistischen Landeskultur in der DDR," January 1969, 3–4.

34 BArch DC 20/19094, "Vortrag auf dem Seminar der Arbeitsgruppe Weimer der Liga für die Vereinten Nationen in der DDR am 25. 9. 1971"; DC 20/21188, "Zur Entwicklung der sozialistischen Landeskultur in den Jahren 1967 bis 1971," September 9, 1971.

35 BArch DC 20/19102, "Die Entwicklung der soz. Landeskultur in der DDR – Ausdruck der kontinuierlichen Politik der SED zur Gestaltung der entwickelten soz. Gesellschaft," January 20, 1971.

36 Huff, *Natur und Industrie im Sozialismus*, 184–185. BArch DK 5/3386, "Bericht über Erfahrungen und Ergebnisse der Arbeit der Staatlichen Umweltinspektion beim Magistrat der Hauptstadt der DDR," June 23, 1987.

37 BArch DK 5/1337, "Bericht über die Ergebnisse der Entwicklung des Umweltschutzes in der DDR, 1976–80."

resources.[38] The inclusion of "water management" in the Ministry for the Environment and Water Management, founded in 1972, indicated that it was a vital but challenging component of the SED's pursuit for a cleaner environment. Water stood at the confluence of the SED's most progressive ideals and the realities of its economic situation.

Water's Connectivity

In the context of the Cold War, and as the strongest economy in the eastern bloc, the SED often touted East German technological innovation and superiority.[39] This framing followed the priorities set out in the ESS, but also aimed to gain the GDR the international recognition it desired.[40] In the 1970s, the SED portrayed technology, environmental protection, and water management as intersecting arenas in which the GDR excelled. While looking to the world beyond the Iron Curtain, the SED also sought to demonstrate the GDR's prominence within the socialist sphere. Titel had pointed to the Soviet Union to prod the SED into greater environmental protection domestically, but to other countries, the GDR presented itself as an environmental and technological leader. The GDR forwarded this reputation within the Soviet bloc, even as it faced ongoing water shortages and pollution at home, highlighting the importance of water to the economic and natural ecology of socialism. The GDR turned to fellow Comecon countries for help as it sought to establish itself as a critical node in that system.

38 "Gesetz über die planmäßige Gestaltung der sozialistischen Landeskultur in der Deutschen Demokratischen Republik – Landeskulturgesetz – vom 14. Mai 1970," *Gesetzblatt DDR* (May 28, 1970).

39 Charles Maier, *Dissolution: The Crisis of Communism and the End of East Germany* (Princeton, NJ: Princeton University Press, 1997), 82.

40 Kai Hünemörder, "Environmental Crisis and Soft Politics: Détente and the Global Environment, 1968–75," in *Environmental Histories of the Cold War*, eds. John R. McNeill and Corinne Unger, 257–276.

In addressing conditions in the GDR in the 1970s, experts emphasized the "necessary multiple uses of water" and finding the means to undertake new projects. To provide for the population, industry, and agriculture, the Office of Water Management built and updated reservoirs in order to provision the population with water. It also upgraded facilities for wastewater treatment as a means of alleviating the pollution from byproducts and runoff, especially in refining brown coal and in the chemical industry.[41] Enterprises had to pay, or compensate the state, for water usage. Indeed, according to a 1970 amendment to the 1963 Water Law, any money raised from fines were "obligated to cover societal costs incurred by not remaining within the legal bounds."[42] Moreover, the penalties for polluting could not be included in an industry's Five Year Plan nor be calculated ahead of time, meaning that the enterprise had to shoulder the burden on its own.[43] After pushing for a reconsideration of water and the environment in the 1960s, the new measures aimed to regulate those considerations.

A report from the newly minted Ministry for Environmental Protection and Water Management (MUW) in 1972 did not want to rush to judgment, but pointed to "limited improvements in specific areas."[44] It acknowledged that the "rise in water pollution had not been hindered," and therefore, argued that to better "the situation, it is essential to secure strict order, discipline, and qualified service" in various industries, factories, and collectives. Horst Wambutt, head of the primary industry department in the environmental ministry, recommended the construction of new facilities, especially on rivers critical to industrial use.[45] While conceding

41 BArch DK 5/1831, Wambutt, "Betreff: Information über den Stand und die Entwicklung des Umweltschutzes in der DDR," April 5, 1972, 4.

42 BArch DC 20/I/3/830, Ministerrat der DDR, "Beschluß zur 2. Durchführungsverordnung zum Wassergesetz über die Anwendung von ökonomischen Regelungen für die Reinhaltung der Gewässer und zur rationellen Nutzung des Grund- und Oberflächenwassers vom 16. Dezember 1970," 1.

43 BArch DC 20/I/3/830, Ministerrat der DDR, "Beschluß zur 2. Durchführungsverordnung zum Wassergesetz," 2.

44 BArch DK 5/1831, Wambutt, "Betreff: Information über den Stand und die Entwicklung des Umweltschutzes in der DDR," April 5, 1972, 8.

45 BArch DK 5/1831, Wambutt, "Umweltschutzes in der DDR," 8.

a lack of significant progress, the report recommended more collaboration between socialist bloc countries to find solutions to these problems. The multilateral cooperation between water management offices and departments that had been in existence since 1961 was to be deepened. A 1971 agreement among Comecon countries formalized the sharing of scientific and technical knowledge and "socio-economic and legal questions" on nature conservation.[46] The GDR's work with other countries was to result in a stronger economy and a cleaner environment.

To that end, the GDR pursued agreements, and eventually a formal accord, within the Soviet bloc that demonstrated its forward thinking and resolved domestic shortcomings. As early as 1972, the idea of the *Interwodootschistka* was developed at a meeting of the Conference of Leaders of Water Management Organizations in the Comecon (TLWO) in Budapest.[47] After more formal approval at another meeting in 1974 in Ulan-Bator, Mongolia, a working group was formed to determine what the *Interwodootschistka* would look like, especially with regard to its responsibilities and economic considerations. Johannes Rochlitzer, East German head of water management, enthusiastically sought to position the GDR advantageously within the debate and eventual organization itself. Rochlitzer reached out to numerous internal institutions, such as the Ministries for Chemical Industry, Coal, and Energy and Foreign Affairs, to garner support, and ultimately, received a preliminary consensus at home.[48]

Rochlitzer then turned to the TLWO's working group to hash out the details. Under his guidance, the GDR's delegation advocated for cooperation and specialization in building and producing water management facilities and equipment as well as a coordination center. The most important considerations were keeping any financial burdens to a minimum while unifying science internationally.[49] For Rochlitzer, increasing the production of wastewater equipment and machinery was necessary for the GDR

46 BArch DK 5/1831, Wambutt, "Umweltschutzes in der DDR," 18–19.

47 BArch DK 5/1046, Schriftwechsel zwischen Dr. Gerhard Weiss and Diplom-Ingeneur Johannes Rochlitzer, 1975.

48 BArch DK 5/1046, Schriftwechsel.

49 BArch DK 5/1046, Schriftwechsel.

(and other Comecon countries), because it represented "a bottleneck in industrial and communal areas" and hemmed in their economies.[50] The *Interwodootschistka*, and the international cooperation it represented, was an essential step in overcoming economic and technological challenges within the communist system, including outdated facilities and a lack of cooperation. Better water conditions and supply, however, would also work to fulfill socialism's promise of providing for its citizens, both materially and spiritually.[51]

Finally, Bulgaria, Hungary, the GDR, Poland, Romania, the Soviet Union, and Czechoslovakia signed the treaty creating the *Interwodootschistka* in Moscow in April 1977. Going into effect a year later, it foregrounded the centrality of water to the functioning of the economy and the technological cooperation imperative to that. It focused on technical issues, such as organizational concerns, the buying and selling of licenses and patents, diagnosing needs, and development of research.[52] Among other tasks, the member countries aimed to share information and technology regarding sewage, the design of water treatment plants, and the construction of machinery for plants and instruments for measuring.[53] Despite specific parameters, the agreement emphasized that water management was not purely for economic and technological reasons. The organization also aimed to provide "a faster and more effective solution to the problem of environmental protection." By working together, the member countries could develop solutions – and technology – to resolve common difficulties.[54] The Comecon and *Interwodootschistka* acknowledged the economic and environmental impact of polluted water.

50 BArch DK 5/1046, Schriftwechsel.

51 Konrad H. Jarausch, "Care and Coercion," in *Dictatorship as Experience: Towards a Socio-cultural history of the GDR*, eds. Konrad H. Jarausch et al. (New York: Berghahn Books, 1999), 57–62.

52 BArch DK 5/2248, Artikel V, Absatz 2, "Abkommen über die Gründung der Internationalen Wirtschaftsgemeinschaft 'Interwodootschistka,'" Moscow, 1978.

53 BArch DK 5/5765, Klaus Heilmann, "Bericht zum Stand der Arbeiten in der Internationalen Wirtschaftsgemeinschaft 'Interwodootschistka,'" September 18, 1980.

54 BArch DK 5/2248, "Abkommen über die Gründung der Internationalen Wirtschaftsgemeinschaft 'Interwodootschistka,'" Moscow, 1978.

The GDR further formalized its international cooperation on water within the Soviet bloc by signing an environmental treaty with Poland in 1973. The relationship with Poland served a practical purpose, being a neighbor with which the GDR shared rivers and Baltic Sea coast, but was also symbolically vital as a fellow Soviet bloc state. The treaty specifically emphasized the centrality of water to the two countries and the exchange of water management technology for environmental purposes. Indeed Poland and the GDR's most prominent water management enterprises, in Halle and Warsaw, respectively, were to deepen their collaboration on ways to recirculate surface water, chemical methods to clean it, and automation of water treatment.[55] These endeavors were to "avoid parallel work on Comecon topics" (e.g. work in the TLWO or *Interwodootschistka*) and to "prioritize the concrete division of labor without significantly [raising] the number of topics" to be addressed.[56] Still, water management dominated Polish-East German environmental diplomacy in the 1970s and 1980s.

East German and Polish representatives laid out joint projects after 1973, including water in their respective Five Year Plans, despite frustrations over rights and payments. In 1978, experts from both countries met to discuss the upcoming 1980 to 1985 period. Representatives proposed to research "new processes for treating heavily polluted surface water into drinking water" and to compare various methods into order to find the most "optimal" one.[57] For industry, too, water rights remained tricky.[58]

55 BArch DK 5/1055, "Liste für Vorschläge zur direkten Zusammenarbeit zwischen Betrieben und Einrichtungen der Wasserwirtschaft mit polnischen Partnerbetrieben," undated; BArch DK 5/530, "Einschätzung der Zusammenarbeit zwischen der DDR und der VRP auf dem Gebiet des Umweltschutzes für das Jahr 1975," undated.

56 BArch DK 5/530, "Direktive für die Expertenberatung zur Vorbereitung der 3. Verhandlung der Regierungsbevollmächtigten zur Durchführung des Umweltschutzabkommens DDR/VRP," undated, 1–2.

57 BArch DK 5/529, G. Thoms and Z. Drozd, "Arbeitsplan über die Zusammenarbeit im Jahre 1978," February 2, 1978.

58 BArch DK 5/1991, Hans Reichelt, "Direktive für das Auftreten auf dem dreiseitigen Treffen der Stellv der Vors des MR der DDR, CSSR und VRP zu Fragen des Umweltschutzes und der Wasserwirtschaft im Februar 1988 in der VRP."

Pollution levels in the two rivers stabilized in the 1970s – thanks to a 1965 agreement – but the use of water for cooling power plants remained tense.[59] Into the 1980s, the GDR paid Poland for water from the Vistula and the Neisse for its plants at Hagenwerder and Hirschfelde, switching entirely to the Neisse. Negotiations stalled, though, because in 1983, Poland demanded a higher price for its water. Within the GDR, then, MUW and the Ministry for Coal and Energy bickered over who should pay it and how much of the price was offset by construction costs that the GDR had already accepted.[60] Water connected Soviet bloc countries through shared problems, communal waters, and international agreements.

The Tide Turns

Water usage and pollution followed the GDR and other Soviet bloc countries into the 1980s, contributing to their economic and environmental difficulties. In the GDR, water quality fell into the lowest category (Potsdam, Dresden, Leipzig, Erfurt, and Karl-Marx-Stadt) in nearly third of all districts and represented well over a third of the population.[61] The MfS took more control of water issues to ensure the functioning of the economy. This transition demonstrated the perceived need for secrecy around the environment, which Council of Ministers' 1982 decision to classify all environmental data confirmed. Hiding data rather than investing heavily in updating and improving technology reflected the party and state's loss of a monopoly over the environment, as did the rise of

59 BArch DK 5/2485, "Konzeption für das Gespräch am 4. Juli 1973 mit Genossen Professor Okuniewski," undated.

60 BArch DK 5/1055, "Bezahlung der Jahrespauschals für die Wasserbereitstellung der Kraftwerke Hirschfelde und Hagenwerder an die VR Polen," August 25, 1983.

61 BArch DK 5/2145, "Bericht über Ergebnisse des Umweltschutzes in der Deutschen Demokratischen Republik, 1981."

environmental groups within the Protestant Church.[62] The relationship among Comecon states changed, too, as environmental cooperation did not yield better water quality or ease waters shortages. In the 1980s, the *Interwodootschistka* even recommended moving away from large-scale, multilateral treaties and instead approaching water questions through narrower bilateral and trilateral agreements.[63]

The MUW's efforts led to some progress in the latter half of the 1970s, though larger problems surrounding water pollution remained, as industry came into conflict with regulation. A 1981 report noted that measures introduced to use water more efficiently resulted in a 5.5 percent reduction in water consumption. Unfortunately, the decrease there was offset by an increase in other industries and agriculture. Similarly, the MUW's policies resulted in lower sewage levels in the surface water of lakes and rivers, but "only attained isolated improvements in [overall] water quality."[64] For example, the Ministry struggled to regulate collective farms or check the flow of nitrates used in pesticides and fertilizers into the groundwater.[65] The MUW nevertheless concluded its goals for the year 1981 were "99.2 percent fulfilled" and that 55.1 percent of the investments had gone into abatement.[66] Officials quantified their successes in obscure terms, using the language of the economic plan to highlight – or exaggerate – accomplishments. Given the MUW's inability to convince high level political and economic leaders – such as Günter Mittag – to heavily invest in water protection, the Ministry's efforts "had not yet led to a fundamental improvement in the

62 BArch DK 5/1982, "Bericht über Probleme des Geheimnisschutzes beim Infationen zum Umweltschutz," October 25, 1982. For more on the rise of environmental groups in the Protestant Church, see M. E. Jones, "Origins of the East German Environmental Movement," *German Studies Review* 16, no. 2 (1993): 235–264.

63 BArch DK 5/5765, Dr. Klaus Heilmann, "Bericht zum Stand der Arbeiten in der Internationalen Wirtschaftsgemeinschaft, Interwodootschistka,'" September 18, 1980.

64 BArch DK 5/2145, "Bericht über die Ergebnisse des Umweltschutzes in der DDR im Jahre 1981," undated.

65 Huff raises similar challenges in regulating air pollution: Huff, *Natur und Industrie im Sozialismus*, 270–273.

66 BArch DK 5/2145, "Bericht über die Ergebnisse des Umweltschutzes in der DDR im Jahre 1981," undated.

[water] situation."[67] In some places, such as Halle, residents boiled water for children and expectant mothers throughout the 1980s.[68]

In 1982, a working group in the MUW proposed a new Water Law in an attempt to correct systemic challenges. The law was passed, and went into effect that year. While maintaining the applicability of the 1963 law, the new law updated the old statute with more draconian measures against polluters. It continued to emphasize the interconnectedness of water protection, economic considerations, and the needs of the population, and to do so, reiterated the more "efficient use" and "reduction in quantity" of water in industry, the "reclamation of resources from wastewater," and other nearly identical phrases that experts had been using since the 1960s.[69] It resolved to "[extend] the permit requirement" for using water above an allotted amount, thereby making it more difficult for industries to ignore their legal limits. It also included sea water more explicitly.[70] The Politburo tacitly acknowledged its water shortages and quality problems, but often remained entrenched in decades-old language about developing more efficient technology without being able to deliver that technology.

The lack of progress – along with growing public awareness about environmental conditions – became a major reason for the SED's decision to classify all relevant data in 1982. The fear that "imperialist forces" in the West "misused" statistics about pollution and health to undermine socialism was a common theme in the arguments for that action. MUW officials went so far as to argue that Western publications revealed data about acid rain levels along the East German-Czechoslovak border in

67 BArch DK 5/2145, "Bericht über die Ergebnisse des Umweltschutzes in der DDR im Jahre 1981," undated. See also Joachim Radkau, *The Age of Ecology: A Global History* (New York: Polity, 2014), 346.

68 Robert Havemann Gesellschaft (RHG) RG/SA 03, "Ökologishes Merkblatt, Umweltgruppe Christusgemeinde Halle/S," undated.

69 BArch SAPMO DY 30/J/IV 2/2/1944/01, "Grundsätze für das neue Wassergesetz," May 4, 1982, 2–3.

70 BArch SAPMO DY 30/J/IV 2/2/1944/01, "Grundsätze für das neue Wassergesetz," 3–4.

order to "sow discord between the friendly relations" of those two countries.[71] The domestic situation also factored into the decision. A growing number of independent environmental groups in the Protestant Church contested the SED's environmental commitment, protesting in the name of the environment. The resolution to classify the data contended that the members of these groups were "class enemies" who "took advantage of … pollution that objectively cannot yet be eliminated."[72] Rather than face the threats, the Council of Ministers hid the data.

Given the sensitive character of environmental data, the MfS began to take a more active role in handling water matters. The MfS had monitored water mishaps since the 1970s, falling under the jurisdiction of Department XVIII, "Administration of the Economy." These MfS generated memos on how to respond to West German inquiries into water pollution, neutralizing chemical spills into rivers that flowed into the FRG, and gauging different industries' requirements. In 1976, a serious chemical spill at an enterprise in Coswig released hydrochloric acid into the Jeetze River, which killed all the fish in the area. The MfS then sent border troops to pull dead fish out of the water before they crossed the border, in order to conceal the extent of the contamination from West German officials.[73] Throughout the 1980s, the MfS reacted to environmental accidents and shaped official narratives about pollution. This division of labor had to be sorted out between the MUW and the MfS in addition to mass social organizations, such as the Cultural League's Society for Nature and Environment (successor to the Friends of Nature and Heimat). At the same time, the GDR negotiated with the West German government over the quality of water that flowed westward and technology to clean it.[74]

71 BArch DK 5/1982, Arbeitsgruppe für Organisation und Inspektion beim Ministerrat, "Bericht über Probleme des Geheimnisschutzes beim Informationen zum Umweltschutz," October 25, 1982.

72 BArch DK 5/1982, "Bericht über Probleme des Geheimnisschutzes."

73 Behörde des Bundesbeauftragten für die Stasi-Unterlagen (BStU), Ministerium für Staatssicherheit (MfS) HA XIX 19386, "Information über Fischsterben in der Jeetze, Krs. Salzwedel/Magdeburg," September 11, 1976.

74 Eckert, "Geteilt, aber nicht unverbunden," 86–93.

An agreement between the MUW and the MfS formalized a joint duty to water and environmental questions in 1985, despite stark differences in priorities. MfS Lieutenant General Schwanitz and others consistently highlighted their ministry's prioritization of water over other environmental concerns. They focused particularly on the "Prevention of water pollution disasters as well as the consistent lowering of water pollution through contaminants," as stated in the 1982 Water Law.[75] Presumably, water quality (and quantity) more directly affected the functioning of the economy than the air or an abstract concept of environmentalism. The MfS also became responsible for limiting exemptions to standards and implementing a "polluter pays" model for East German industries.[76] Even the MfS admitted that "comprehensive compliance with the Water Law demanded (besides a range of leadership-oriented measures) substantial investment ... that in the short term" was not possible.[77] The MfS's goal remained the security of the state. The economy was a critical component of maintaining stability, but environmental protest – and an independent movement growing in the Protestant Church – threatened it.

The MfS also dove into how water management was presented abroad, specifically working with the MUW on treaties and agreements in the 1980s. Indeed, rather than shutting down as the GDR faced political challenges and shifting power within the Soviet bloc, the two ministries pursued new international avenues. The MUW reported to the MfS on water management collaboration in the Comecon, complaining that "work [in the *Interwodootschistka*] is too ponderous and longwinded. Therefore, there will be no real results in the short term." The MUW official who lodged this complaint believe that the lack of efficacy was because "the

75 BStU MfS VRD 8709, Arbeitsgruppe beim Stellvertreter des Ministers, "Bericht zu den Ergebnissen der Überprüfung zum Stand der Durchsetzung der VVS 74/86," November 2, 1987.

76 BStU MfS VRD 8709, "Auskunftsbericht über die Durchsetzung der Erfordernisse des Gewässer- und Umweltschutzes im Verantwortungsbereich des Ministeriums für Staatssicherheit," October 20, 1989.

77 BStU MfS VRD 8709, "Meinungsäußerung zu den vorgeschlagenen Maßnahmen zur Verbesserung des Gewässer- und Umweltschutzes," March 20, 1989.

participants from different countries in the society are equipped with different powers," meaning that some had more domestic authority than others.[78] Moreover, the enterprises building equipment for water treatment plants managed to cover domestic needs, but did not produce enough to fulfill its Comecon commitments. In particular, the delivery of specialized machinery was inconsistent.[79] These holdups led the technical director of the enterprise in Halle to request permission to "no longer participate" in the *Interwodootschistka*.[80] Despite ongoing efforts in the 1980s, international water management and environmental protection made more progress outside of the Comecon, in the domain of bilateral or trilateral treaties.

The GDR advanced its status as a leader of environmental protection and technological innovation, despite the eastern bloc's shifting political landscape with glasnost and perestroika. In 1988, the GDR and the Soviet Union signed a treaty on environmental protection. In MfS directives on how East German delegates should act, Stasi officials smugly noted that the resolution for the treaty "fully met the GDR's interests" and that it "reflected proposals that the GDR had repeatedly presented in recent years."[81] The treaty in May swiftly followed the establishment of a Soviet environmental ministry in February of 1988. East German officials hoped, "With this treaty, there will be the opportunity to add to the collaboration of urgent tasks regarding the economy and water management as pertains to the protection of nature and efficient use of natural resources."[82] Rather than giving up amid perestroika and economic turbulence, the two countries renewed endeavors to exchange technological innovation "to raise the

78 BStU MfS HA XVIII 26456, "Fragen der internationalen Zusammenarbeit im RGW," June 10, 1987.

79 BArch DK 5/5765, "Information zu Punkt 13 der vorläufigen Tagesordnung der 29. TLWO-Sitzung der Mitgliedsländer des RGW 'Über die Tätigkeit der Internationalen Wirtschaftsgemeinschaft Interwodootschistka,'" undated.

80 BArch DK 5/5765, Klaus Heilmann, "Bericht zum Stand der Arbeiten in der Internationalen Wirtschaftsgemeinschaft 'Interwodootschistka," September 18, 1980.

81 BStU MfS Rechtsstelle 210, "Direktive für das Auftreten der DDR-Delegation in den Verhandlungen mit der UdSSR zur Ausarbeitung eines Regierungsabkommens über die Zusammenarbeit auf dem Gebiet des Umweltschutzes," May 4, 1988.

82 BStU MfS Rechtsstelle 210, "Direktive," May 4, 1988.

efficiency of the economy and to improve the working and living conditions of the populations."[83] The directives and official treaty between the Soviet Union and the GDR explicitly acknowledged ecological stumbling blocks, seeking "new" solutions despite a lack of innovation and means.

In the final years of the Soviet control over eastern Europe, the GDR and its communist neighbors collaborated more closely on water and shared environmental issues. The GDR pushed for the creation of a trilateral treaty with Poland and Czechoslovakia in 1988 that addressed polluted rivers and air, especially in the Sudeten, where the three countries bordered one another.[84] Yet such an arrangement posed potential challenges. East German air pollution disproportionately affected Poland (and to a lesser extent Czechoslovakia), but polluted water in Czechoslovak rivers overwhelmingly had an impact on the GDR.[85] Thus while Poland pursued concrete obligations for reducing emissions, the GDR emphasized the importance of water conditions that had an impact on its industry along the Oder and Neisse Rivers. Within these discussions, the Czechoslovak Prime Minister Ladislav Adamec remained relatively quiet.[86] Setting standards meant to ensure accountability was promising but nevertheless quick to fall into old patterns, in which negotiations over technicalities left larger issues unresolved.

Meetings between these three countries were fraught as independent organizations and the press further weighed in. Poland, in particular, loosened censorship over the press in the final years of communism, permitting experts and journalists to criticize the government. Calling the trilateral treaty "the first agreement of its kind between socialist states," the Polish press praised it as "a model for further ecological collaboration in our

83 BStU MfS Rechtsstelle 210, "Direktive," May 4, 1988.

84 Eagle Glassheim addresses the 1980s in *Cleansing the Czechoslovak Borderlands*, 162–176.

85 BArch DK 5/1997, "Direktive für die DDR-Delegation zur Durchführung der 22. Tagung des Ausschusses für die wirtschaftliche und w-t Zusammenarbeit zwischen der DDR und der VRP (Wirtschaftsausschuß)," September 6, 1988.

86 BArch DK 5/1997, "Bericht über die Konferenz der für Fragen des Umweltschutzes verantwortlichen Mitglieder der Regierungen der CSSR und ihrer Nachbarländer vom 29. Und 30. Mai [1989] in Prag," June 21, 1989.

camp."[87] Critical articles in the *Trybuna Ludu*, complained that these talks and any resulting treaty would be non-binding.[88] Polish environmental minister, Józef Kozioł, reported to his East German counterpart, Hans Reichelt, "the ecological consciousness of the Polish population is rapidly increasing," thanks to better information and freer speech.[89] Leveraging this pressure, Kozioł attempted to extract more concessions from the East Germans regarding pollution abatement. The GDR, however, was unmoved and rejected Polish demands for a polluter pays model, because the GDR refused liability for pollution or damages.[90]

A trilateral treaty signed on July 1, 1989, in Wrocław, Poland, concluded the negotiations, but its efficacy would go untested in light of political transformations already underway. Willi Stoph, Chairman of the Council of Ministers, declared the agreement to be "in accordance with the Warsaw Pact" in terms of economic and diplomatic cooperation between Soviet bloc countries.[91] The agreement affirmed the cross-border character of environmental pollution and that each partner held a high degree of responsibility for the protection of natural resources. It further resolved to "improve environmental conditions in the interest of the people's health and the dynamic development of the economy in the three countries," and set unified procedures and regulations for an array of environmental topics.[92] Still many of them, and

87 BStU MfS HA XVIII 25283, "Veröffentlichung der 'Trybuna Ludu' vom 6.9.88, 'Wir müssen die Sudeten retten – Vor einer weiteren Runde der dreiseitige Gespräche,'" undated.

88 BStU MfS HA XVIII 25283, "Veröffentlichung der 'Trybuna Ludu'."

89 BStU MfS HA XVIII 25283, "Notiz über das am 13. August zwischen den Stellvertretern des Vorsitzenden des Ministerrates der DDR, Dr. H. Reichelt, und der VR Polen, J. Koziol, in Warschau geführte Gespräch," undated.

90 BStU MfS HA XVIII 25283, "Information über die 2. Beratung zur Ausarbeitung eines dreiseitigen Abkommens über die Zusammenarbeit auf dem Gebiet des Umweltschutzes der Deutschen Demokratischen Republik, der Tschechoslowakischen Sozialistischen Republik und der Volksrepublik Polen," December 5, 1988.

91 Wolfgang Rex, "DDR, Polen und ČSSR schlossen Umweltschutzabkommen ab," *Neues Deutschland*, July 3, 1989.

92 "Abkommen zwischen der Regierung der Deutschen Demokratischen Republik, der Regierung der Volksrepublik Polen und der Regierung der Tschechoslowakischen Sozialistischen Republik über die Zusammenarbeit auf dem Gebiet des Umweltschutzes

including the language used to discuss them, were nearly identical to what the GDR had been using since the 1960s. *Neues Deutschland*, the SED newspaper, reported on the event, highlighting the mutual "development of low waste and water saving technologies including binding steps for the reclamation and reuse of materials."[93] This final treaty reflected a consistent approach to the environment and water, emphasizing their crucial role economically and socio-politically in socialism. Yet the trilateral treaty also underscored the inability to demonstrably clean up the water or air.

Conclusion

Water flowed through both the GDR's economic and environmental policies, being central to both industrial production and the wellbeing of the East German population. From early on, the SED had to manage water in order to provide for East Germans and grow its economy. Starting the 1960s, the effects of water pollution, loss of byproducts, and reclamation of resources from wastewater were an important part of these discussions. In a period of relative stability and optimism, the SED advanced concepts such as *Landeskultur* and socialist environmentalism to provide both materially and spiritually for the workers in a workers' state. It then developed legislation and institutions to support these projects. Blending pragmatism with idealism, the SED endeavored to make the GDR economically stronger and to fashion it as an environmental role model in order to gain international recognition.

---

vom 1.7.1989," *Gesetzblatt DDR Teil II No. 2* (February 9, 1990), 9–12; Reinhard Müller, Birgit Süß, "Die Entwicklung des Umweltvertragsrechts zwischen Deutschland, der früheren Tschechoslowakei und Polen – unter rückschauender Einbeziehung der Vertragspraxis der DDR," *Zeitschrift für ausländisches öffentliches Recht und Völkerrecht*, 53, no. 2 (1993): 304, accessed May 25, 2017, <http://www.zaoerv.de>.

93 Wolfgang Rex, "DDR, Polen und ČSSR schlossen Umweltschutzabkommen ab," *Neues Deutschland*, July 3, 1989.

East German water management and environmental protection were not only geared toward the international stage, but also specifically within the Soviet bloc and the Comecon. The research and exchange of technology fit within the Soviet bloc's efforts to improve from within. Again, the GDR was crucial in developing these relationships, as the creation of the *Interwodootschistka* highlights. At times, the SED pushed other countries in the bloc, including the Soviet Union, to greater coordination and collaboration. Yet, as the eastern bloc faced stalling economies and growing debt in the 1970s and into the 1980s – such that large investments in water treatment and protection became unfeasible – member countries pursued alternative paths to securing water protection. They made agreements among themselves, strengthened regional ties, and reached out to non-Comecon countries to achieve their own goals. Water's centrality to the functioning of Soviet-style economies with their emphasis on water-intensive industries and "real existing socialism" did not change, though the methods to protect and conserve it did.

If ecology is the interaction between different organisms, then the Soviet ecology of environmental protection was complex, indeed. Power and policy within the Soviet sphere did not always flow from Moscow outward. East Germans pushed the Soviets on various policies from the Berlin Wall to technological and environmental cooperation. The SED struggled to ensure the GDR's prominence within the Soviet bloc in terms of water management and environmental protection, but also pressed other countries, especially neighboring Poland and Czechoslovakia, into stricter measures. A sign of the declining system loomed large when, in the late 1980s, the SED refused to commit to binding steps to reduce the GDR's industrial pollution in the Sudeten. Thus, the GDR's environmental policies and impotence to fully address its own failings left a troubling legacy when the SED lost power in 1989–90. The relationship between the Soviet Union, the GDR, and other Comecon countries functions as an "ecology" on multiple levels. Water use issues formed one of the main ways in which the countries of the bloc interacted with each other, because they all depended on water. As member states formed a network of interdependent relationships based on economic and environmental issues, a political ecology came to mirror a "real" ecology, one in which water was a fundamental underpinning.

THOMAS FLEISCHMAN

The Half-Life of State Socialism: What Radioactive Wild Boars Tell Us About the Environmental History of Reunified Germany

Before 2010, the idea that wild animals regularly attacked human residents of the city of Berlin – the powerful, cosmopolitan metropolis of twenty-first-century Europe – would have seemed "almost incomprehensible" to any German. Yet tell that to the Berliners who were overrun by a sounder of wild boars on Kurt Schumacher Platz near Tegel airport in 2017;[1] or the retirees in Charlottenburg who found themselves on the business end of a particularly sharp-toothed, razor-tusked wild boar in 2012. In the latter case, a 24-year-old good Samaritan attempted to run off the boar, but instead found himself trapped and bleeding on the hood of a parked car. Only the arrival of police, and a service revolver, brought the animal's rampage to an end. In the former case, state hunters shot a boar in nearby Rehberge park. It was unclear if the 18 shots ended the life of the criminal offender, or an innocent swinish bystander, retribution meted out as a warning to others.[2]

Wild boars have been uprooting gardens, destroying patios, and devouring farm harvests in Germany for years. The problem, however, has become more sensational in the last decade as encounters between humans and pigs have driven newspaper headlines everywhere: boars have knocked

1 Katharina Metag, "Keiler-Angriff in Rehberge, Wildschwein mit 18 Schusse getötet," *Berliner Zeitung*, January 28, 2017, accessed June 20, 2017, <http://www.bz-berlin.de/berlin/mitte/keiler-angriff-in-rehberge-wildschwein-mit-18-schuss-getoetet>.

2 Charles Hawley, "Radioactive Boar on the Rise in Germany," *Der Spiegel*, July 30, 2010, accessed June 4, 2017, <http://www.spiegel.de/international/zeitgeist/a-quarter-century-after-chernobyl-radioactive-boar-on-the-rise-in-germany-a-709345.html>.

old women from bicycles, assaulted the wheelchair-bound, desecrated cemeteries, and to the horror of this dog-loving nation, severely wounded family pets.[3] Across Germany, authorities have organized special hunting brigades to cull overpopulated areas. But like a zombie plague, wild boar numbers only continue to grow. In 1991, the German boar population sat at about 400,000. Today it has risen to around 2.5 million. The boar plague has even grown to international proportions. Similar stories can be found in newspapers from Texas to Japan.

Of all the sensational headlines, however, none has raised as much curiosity and fear as the discovery of radioactive boars. Newspaper reports first surfaced around 2010, but picked up in intensity after the Fukushima nuclear disaster in 2011. After a 2012 law mandated the testing of wild game in Germany, forestry officials confirmed multiple radioactive cases in the woods of Saxony in the east and Bavaria and Baden in the south.[4] Radioactive boars have also been found further afield in Switzerland, France, Northern Italy, Austria, and Croatia. The most likely explanation, of course,

3 "Wildgewordener Keiler greift Fahrradfahrer an," *Frankfurter Neue Presse*, March 11, 2013, accessed June 20, 2017, <http://www.fnp.de/rhein-main/Wildgewordener-Keiler-greift-Fahrradfahrer-an;art801,478702>; Charles Hawley, "Radioactive Boar on the Rise in Germany," *Der Spiegel*, July 30, 2010, accessed June 4, 2017, <http://www.spiegel.de/international/zeitgeist/a-quarter-century-after-chernobyl-radioactive-boar-on-the-rise-in-germany-a-709345.html>; Rainer W. During, "Wildschweinplage auf dem Friedhof dauert an: Erneut Gräber verwüstet," *Der Tagespiegel*, October 4, 2016, accessed June 20, 2017, <http://www.tagesspiegel.de/berlin/bezirke/spandau/wildschweinplage-auf-dem-friedhof-dauert-an-erneut-graeber-verwuestet/14640798.html>; Kenny Langer und Annechristin Bonß, "Hund von Wildschwein angegriffen," *Sächsische Zeitung*, February 16, 2017, accessed June 20, 2017, <http://www.sz-online.de/nachrichten/hund-von-wildschweinen-angegriffen-3614162.html>.

4 "Tausende Radioaktive Wildschweine in Bayern," *Frankfurter Allgemeine Zeitung*, March 15, 2013, accessed June 23, 2017, <http://www.faz.net/aktuell/gesellschaft/tschernobyl-folgen-radioaktive-wildschweine-in-bayern-13484151.html>; "Strahlende Wildschweine in Sachsen: Was ist in unseren Wäldern noch alles radioaktiv?" *Bild Online*, September 1, 2014, accessed June 23, 2017, <http://www.bild.de/news/inland/wildschweinfleisch/strahlende-wildschweine-was-ist-noch-radioaktiv-im-wald-belastet-37487136.bild.html>.

points back to Chernobyl. In April of 1986, an explosion at reactor number four sent a plume of radioactive fallout across Europe. As the clouds drifted hundreds of miles west, isotopes like cesium-137, mixed with water molecules in rain clouds amassing around the rocky uplands and spring-time cold fronts of the Alps and Hartz mountains. The resulting precipitation produced rain drops laced with cesium-137, which poured down across the watershed. The isotopes drifted first onto the needles and leaves of trees, and then worked their way down to the forest floor. The cesium that made it all the way to the ground moved slowly, centimeter by centimeter, year after year, down into the soil. Once there it became absorbed by vascular networks of fungi, and then consumed by mushroom-loving boars.

Even as cesium persists in the European environment, the dangers posed are less clear. Radioactive isotopes accumulate unevenly in bodies and environments, making it difficult to establish hard limits.[5] Health ministries in Germany have reported radiocesium in wild game at wildly different levels from a high of 6,000 Becquerel/kg in wild boars to 1,000 and 400 Bq/kg for mushrooms and roe deer respectively.[6] Furthermore, the overall number of radioactive boars in Germany remains around 2 percent.[7] And when you also consider the ecological half-life of cesium-137 is "only" 40 years, this poison will pass from the European landscape shortly

5 For an extensive discussion of the unpredictable movement of radioactive isotopes and the variability of bio-accumulation in animal bodies, see Kate Brown, *Plutopia: Nuclear Families, Atomic Cities, and the Great Soviet and American Plutonium Disasters* (New York: Oxford University Press, 2013).

6 "Radioaktive Belastung von Pilzen und Wildbret," Bundesamt für Strahlenschutz. <http://www.bfs.de/DE/themen/ion/umwelt/lebensmittel/pilze-wildbret/pilze-wildbret.html>; Guilford, "Radioactive Boars." The all-time highest number ever measured is 65,000bq/kg, found in a wild boar in a Bavarian forest in 1998, "Die Kontamination von Lebensmitteln nach der Reaktorkatastrophe von Tschernobyl," Bundesamt für Strahlenschutz, p. 3, <http://www.bfs.de/SharedDocs/Downloads/BfS/DE/broschueren/ion/stth-lebensmittel.pdf;jsessionid=5C8CD7CDD976EC1E75AB9860891A440F.1_cid349?__blob=publicationFile&v=6>.

7 Juergen Batz, "Germany's Radioactive Boars a Legacy of Chernobyl," *Associated Press*, April 1, 2011, <https://phys.org/news/2011-04-germany-radioactive-boars-legacy-chernobyl.html>.

too.[8] While radioactive wild boars reveal the on-going, invisible legacy of nuclear energy regimes, they do not pose an existential threat to human life.

While sensational, attacked evening walkers and sickened hunters will not bring the Berlin Republic to its knees. Wild boars, however, can tell us a lot about the politics of post-reunification Germany. The dual "crises" of overpopulation and radioactivity peel back layers of memory and history. Boar overpopulation remains largely understood through presentist concerns over global warming and environmental change.[9] Marauding wild boars become examples of nature out of balance, bad behavior unleashed by a warming planet. Radioactivity, on the other hand, plucks a string resonating all the way back to Chernobyl and West Germany's anti-nuclear history. It reminds people of the Cold War's poor environmental record. The former is rooted in recent global history; the latter in the historical memory of divided Germany.

Nowhere is this clearer than when we look at the wild boar crises through the larger context of Germany's recent environmental record, and in particular through the lens of the *Energiewende*. The sudden "discovery" in 2011 of radioactive boars occurred amidst much larger and renewed angst over nuclear energy. Spurred on by the Fukushima disaster in March of 2011, the German government announced plans to shutter all eight of the country's nuclear power plants by 2022. Furthermore, it set the goal

8 Ecological half-life is the amount of time it takes for an isotope's radioactivity to be reduced in half, by both natural decay and environmental dispersal. Recent studies at Chernobyl, however, have suggested that cesium-137's ecological half-life may be considerably longer, perhaps as long as 180 years. Scientists offer several, and as of yet unproven, explanations. It may be that Chernobyl's marshy soils hold on to cesium longer than expected, or new isotopes continue to work their way up from deep in the earth. The new data, however, has not been enough to overturn the existing number of 30–40 years. Alexis Madrigal, "Chernobyl Exclusion Zone Radioactive Longer than Expected," *Wired Magazine*, December 15, 2009, <https://www.wired.com/2009/12/chernobyl-soil/>.

9 David Crossland, "Climate Change's Clear Winners: Europe's Wild Boar Population," in *Der Spiegel Online International*, November 25, 2009, accessed June 16, 2014, <http://www.spiegel.de/international/zeitgeist/climate-change-s-clear-winners-europe-s-wild-boar-population-exploding-a-663411.html>.

of producing 80–90 percent of its energy needs through renewables by 2050 – an unprecedented benchmark for one of the world's largest economies. The ambitious energy transition, while understood as abrupt, was actually built upon decades of environmentalist protest, organization, and finally legislation. Proponents quickly pointed out Federal legislation, like the Renewable Energy Act in 2000 and the Feed-in Act of 1991, laid the groundwork for implementation of the *Energiewende*.[10] More broadly, the nuclear phase-out fit within Germany's growing reputation as *the* global environmental leader – the "Greenest Nation."[11]

The dual boar crises, however, suggest this conventional wisdom overlooks one major actor in this history, East Germany. For one, Germany's green reputation was burnished during the decade in which the GDR ceased to exist. The laws and treaties that rolled up the state socialist economy and society leveraged environmental pollution and clean-up against the demands for autonomy and work of former East Germans. Instead the GDR was swept aside in the triumphalist narrative of *Die Wende*, or transition to reunification. In popular memory, the collapse and disposal of the GDR was speedy and tidy; it resulted in the "freeing" of millions from totalitarian rule; and in its place the *Wende* gave East Germans a cleaner environment, better food, and civil liberties. The rush to reunification – to reform, to wipe away East Germany – also meant a rush to forget. The myth of the *Wende* elided the environmental and economic legacy of the former states of the GDR. And when the Berlin Republic chose to deploy the term again in the form of the "Energiewende," it was recalling the old myth in the service of a new one – a hope that Germany's energy *Wende* will be as speedy and utopian as the myth of the political *Wende* too.

The *Energiewende*, however, has already come face to face with the complex communist past. An environmental history of post-socialism suggests the GDR continues to exert an influence on the ecology and economy of the Berlin Republic. It refuses to pass into obscurity. Wild boars, the

10 Craig Morris and Arne Jungjohann, *Energy Democracy: Germany's Energiewende to Renewables* (London: Palgrave Macmillan, 2016), 3.

11 Frank Uekötter, *The Greenest Nation? A New History of German Environmentalism* (Cambridge, MA: MIT Press, 2014).

formally collectivized landscapes they lived in, and the radioactive isotopes that laced their skin, tissue, and cells were all products of the socialist states that once ruled central and eastern Europe. In the era of the *Energiewende* Germans are debating questions once raised about the environment and economy of East Germany in the 1980s and 1990s. What is the proper role of brown coal in a modern economy? What are the costs of nuclear power? What is the proper balance between environmental protection and economic growth? Can industrial agricultural follow sustainable practices? How should the state manage destructive wild boars? At the moment in which radiocesium begins to fade from the European environment, German historians can now answer another question: what is the historical half-life of state socialism? Approximately 30 years.

The German Environment, from *Wende* to *Energiewende*

Although two decades passed between *Wende* and *Energiewende*, a number of striking continuities reveal how the politics of reunification exert a powerful influence over the imagined spaces of post-socialist Germany. The term "Wende" resonate across several historical events in German history. Before 1990, German politicians attached the term to end of the social-liberal governing coalition in 1982 and the succession of the center right alliance.[12] Going even further back, anti-nuclear activists coined the latter to describe a democratic and environmentalist plan to reform Germany's energy sector in 1980.[13] For most Germans, however, *Wende* acquired widespread usage following the collapse of the Berlin Wall in 1989 and the sudden rupture it opened in German politics. Into the void, journalists,

12 Charles Maier, "'Als Wär's ein Stück von uns …': German Politics and Society Traverses Twenty Years of United Germany," in *From the Bonn to the Berlin Republic: Germany at the Twentieth Anniversary of Unification*, eds. Jeffrey J. Anderson and Eric Langenbacher (New York: Berghahn Books, 2010).

13 Morris and Jungjohnann, *Energy Democracy*, 4, 29–31.

politicians, and citizens poured their aspirations, plans, and prejudices. By the mid-1990s, popular memory of the *Wende* had codified around a set of precepts regarding democratic rights, neo-liberal economic reform, and environmental protection in the lands of the former East. Implicit in this utopian vision was a wholesale condemnation of the GDR; the result was a full-scale unwinding of the laws, institutions, and traditions of the former communist state.

Admittedly both East and West Germans found many sins for which the GDR had to atone in 1990. First and foremost were the crimes of the Stasi and the regime, particularly political censorship, surveillance, and the murder of East Germans escapees. Things became more contentious, however, when it came to economic reform surrounding privatization and environmental pollution. While there was agreement that reform was needed, the course of that reform was still undecided in 1990. Over the succeeding years, western politicians and media coopted the widely held environmental critiques of the GDR to dominate the course of the *Wende*.

Western media reserved some of its harshest language for industrial pollution. As Raymond Dominick shows, the national magazine *Der Spiegel* derided the East German state as an "ecological outlaw of the first rank." Meanwhile *The New York Times*, homing in on environmental chaos, claimed that the "one issue taking on urgency is how the orderly and clean half of the country can help clean up the disheveled and polluted half."[14] In this way, politicians used western environmental views to delegitimize East German interests from the future of reunification. Instead they urged the GDR's complete economic dismantling. This was especially true in the case of the country's energy and agricultural sectors.

In the last year's of the regime, air pollution had become the most urgent threat to East Germany's public health and environment, and the country's power plants were largely to blame. Since the late 1970s, lignite, also known as brown coal, had become the East Germany's primary source of energy. Lignite is poor fuel. Its energy content is low and therefore burns inefficiently, leaving a high percentage of non-combustible material behind.

14 Raymond Dominick, "Capitalism, Communism, and Environmental Protection: Lessons from the German Experience," *Environmental History* 3, no. 3 (1998): 311.

And what does combust produces toxic dust particles in the form of sulfur dioxide, which blackens concrete and glass, poisons trees and branches, and clings to skin and lungs.[15] Brown coal is nasty, but it is also cheap. And the GDR had a lot of it. Against the background of a twinned debt and energy crisis, the GDR was forced to rely upon the inexpensive and readily available fuel. By 1989 the GDR consumed more brown coal than any other country in the world, burning approximately 40 percent of the entire global supply.[16]

The consequences of brown coal production were considerable. The greatest concentration of sulfur dioxide pollution fell in the industrial south, concentrating around the cities of Leipzig, Karl-Marx-Stadt, and Cottbus. In the city of Halle, scientists estimated that East Germans lost an average of five years of life expectancy compared to the rest of the country. A mere 50 percent cut in particulate emissions, argued health officials, could reduce heart disease by 15 percent, increase life expectancy by three to four years, and decrease respiratory illnesses in the long run by 25 percent.[17] The regime, however, could afford no cuts.

Lignite pollution was not just an East German problem. The Federal Republic had its own brown coal factories, the first generation of which had powered the factories and cities of the *Wirtschaftswunder* of the 1950s and 1960s. Surging economic production brought surging pollution. Between 1962 and 1972, sulfur dioxide emissions doubled across West Germany. Even after the "green decade" of 1980, West Germany's brown coal factories still powered the country. The power plant at Offleben/Helmstedt, which bordered the GDR and later would be switched over the East German net during the *Wende*, emitted sulfur dioxide at 300 parts per billion, nearly

15 Ulrich Petschow et al., *Umweltreport DDR : Bilanz der Zerstörung, Kosten der Sanierung, Strategien für den ökologischen Umbau: Eine Studie des Instituts für Ökologische Wirtschaftsforschung* (Frankfurt/Main: S. Fischer, 1990), 26–27.

16 Uekötter, *Greenest Nation*, 133.

17 Jean DeBardeleben, "'The Future Has Already Begun': Environmental Damage and Protection in the GDR," in *The Quality of Life in the German Democratic Republic: Changes and Developments in a State Socialist Society*, eds. Marilyn Rueschemeyer and Christiane Lemke (Armonk, NY: M. E. Sharpe, 1989), 147.

four times the current limit in the US. Yet this number paled in comparison to East German plants, which produced between three and 10 times the level.[18]

In response to the shared experience of brown coal pollution, West and East German environmental activists collaborated over the course of the 1980s. As is well known, the East German protests of 1989 grew out of a nascent environmental movement at the beginning of the decade.[19] At that time, the *cause célèbre* for environmentalists on both sides of the Iron Curtain was acid rain-induced *Waldsterben* or "forest dieback." In the FRG, *Waldsterben* transformed environmentalist action into political change, as the federal bureaucracy passed major laws regulating power plant emissions in 1983. That same year also saw the Green Party pass the 5 percent threshold in national elections and win seats in the Reichstag for the first time.[20] Acid rain storms were even worse in East Germany, where reportedly 90 percent of the country's forests were affected, including more than 500,000 hectares of standing dead trees, or between 15–20 percent of the country's forests.[21] The transnational threat of acid threat engendered cross-border collaboration. The most famous case began on June 25, 1988,

18 Petschow et al., *Umweltreport, 25*. For this number I converted Petschow et al.'s provided number of 800 mg/m^3 to parts per billion, using the molecular weight of 65.066 g/mol and placing them in a conversion calculator provided here: <https://www.markes.com/Resources/Frequently-asked-questions/How-do-I-convert-units.aspx>.

19 Nathan Stoltzfus, "Public Space and the Dynamics of Environmental Action: Green Protest in the German Democratic Republic," *Archiv für Sozialgeschichte* 43, no. 2 (March 2003): 390; Scott Moranda, *The People's Own Landscape : Nature, Tourism, and Dictatorship in East Germany, Social History, Popular Culture, and Politics in Germany* (Ann Arbor: The University of Michigan Press, 2014). The revolution of 1989 found its roots in the environmental protest movements housed in the Protestant churches of East Germany. East Germans, already heir to tradition of nature conservation, transformed their environmental thought into "western-styled" environmentalist action in the 1980s. Building on the peace movement of the 1970s, East German activists used environmental issues to expand the public space to critique the GDR regime.

20 Uekötter, *Greenest Nation*, 114.

21 DeBardeleben, "'The Future Has Already Begun'," 147.

when a guerilla documentary crew managed to film and then sneak out of the GDR devastating evidence of sulfur dioxide pollution in the industrial south. The footage found its way to producers at a West German television station, who cut it together into a documentary entitled, *Bitteres aus Bitterfeld* or "Bitter News from Bitterfeld." Provocatively, the documentary was broadcast on a West Berlin station, easily finding an audience both in western and eastern homes.[22]

Brown coal pollution was not the only major source of environmental damage in the GDR. Agriculture had also become a significant source of unrest, although its environmental problems were even older than the ones caused by lignite. Beginning the late 1960s, the East German regime promoted the full industrialization of the country's farms. Building on the reforms of collectivization, planners implemented industrial technologies and techniques on both grain farms and livestock producers. Vertical integration, economies of scale, mechanization, agro-chemical fertilizer and pesticides all became common place. Over the course of the 1960s and 1970s, state planners consolidated hundreds of thousands of hectares while introducing unprecedented degrees of industrial organization. They created enormous state-owned farms and livestock CAFOs. Industrialization culminated in the full specialization of each collective farm through the complete separation of grain and animal farming in 1976, in what became known as the Grüneberg Plan. Each newly specialized grain farm managed between 4,000 to 6,000 hectares, and livestock units raised tens of thousands of pigs, cattle, and chickens each.[23] While East German fields swelled in size to 38 hectares on average, their competitors in the Federal Republic remained at about 6 hectares.[24]

Industrialization had mixed results for the GDR. Take the example of meat production. On the one hand, East Germany doubled its output

22 Konrad H. Jarausch, *The Rush to German Unity* (New York: Oxford University Press, 1994), 37; Uekötter, *Greenest Nation*, 136.

23 Hans C. Buechler and Judith-Maria Buechler, *Contesting Agriculture : Cooperativism and Privatization in the New Eastern Germany*, Suny Series in the Anthropology of Work (Albany: State University of New York Press, 2002), 50.

24 Petschow et al., *Umweltreport*, 66.

of pork between 1970 and 1982, becoming the top producer in the eastern bloc, surpassing the Netherlands and rivaling much larger western competitors like the FRG and France.[25] On the other hand, extreme specialization wrought havoc on the environment. Nitrates contaminated drinking water and recreational bodies of water, unleashing a public health crisis. A confidential report from the Ministry of Environment and Water Protection revealed that in 1978, more than 2.75 million East Germans (in a country of roughly 17 million) were exposed regularly to nitrate levels in their drinking water that exceeded the then internationally recognized level of 40mg/liter – a 1,200 percent increase in just 10 years.[26]

Drinking water contamination was not only deadly but also incredibly expensive. As the problem spread into more water tables and fresh water sources, more water wells had to be closed and replaced at a considerable cost to the government. In 1979, the drilling of new wells and installation of new pipes had cost 3.3 million marks just in the district of New Brandenburg. In the same year, the government had provided clean water for infants in 263 villages in the District of Gera, and some 5,000 more infants in the District of Dresden – at a yet to be determined cost. In the area surrounding the city of Cottbus, the regime provided an additional 170,000 bottles of water for families annually, and supplied drinking water for babies in an additional 26 villages. When tallied together, the Ministry for Environmental Protection estimated that the special water deliveries

25 Source, "Production, Supply and Distribution Online," Foreign Agricultural Service, United States Department of Agriculture. International Pork Production Statistics, accessed June 25, 2017, <http://www.fas.usda.gov/psdonline/psdQuery.aspx>. According to the FAO numbers, the GDR increased production from 585,000 metric tons of pork in 1970 to 1.36 million tons by 1981. The Netherlands reported increasing their stocks from 672,000 to 1.149 million over the same period. France and the FRG increased production modestly, reporting 1.64 tons and 2.691 tons in 1981 respectively.

26 DK 5/1730, 6. Ministry of Environmental Protection and Water Management. Nitrate and Water Contamination (1979). Institute of Water Management, "Studie zur Stickstoffbelastung der Gewässer der DDR, ihre Entwicklung in den letzten Jahren und Hauptrichtungen für deren Veränderung," Berlin, February 1979. The internationally recognized limit has since been lowered to 10mg/Liter today.

had cost the regime 250 times the annual cost for the maintenance of East Germany's entire drinking water system.[27] The problem was so bad that even the regime's own ministers could not shade the truth. As the East German Ministry of Environmental Protection and Water Management declared, "Yes, this problem, like the literature shows, has become an increasingly international concern. But in the GDR the danger of nitrate contamination, through the combination of our strained water supply, and dense development of grain and animal production, has become more critical than in any other European country."[28]

Blooming Landscapes and Environmental Reconstruction

Protest against environmental pollution did not cause the collapse of the East German regime. But it did play a major role in shaping perception of the GDR's farms and factories after 1990. During the *Wende*, reformers latched onto the popularity of environmentalism, fashioning its discourse to their own broader critiques of East German society, economics, and institutions. At the same moment Germans were remaking the future of eastern Germany, their leaders were promoting their own international reputation for environmentalism. In 1992, German representatives to the United Nations Rio Earth Summit grabbed the mantel of global leader on climate change and greenhouse gases.[29] In this way Germany's growing global environmental reputation provided cover for more complicated environmental reform in its post-socialist lands.

Environmentalism occupied a primary position in debates during the *Wende*. Domestically politicians directed their environmental ardor against the soft target of the GDR, even though many of the same environmental problems existed in the FRG as well. By eliding the difference between

27 "Studie zur Stickstoffbelastung," 60.
28 "Studie zur Stickstoffbelastung," 117.
29 Uekötter, *Greenest Nation*, 2.

environmental and economic questions, reformers politicized environmental problems in the name of neoliberal economic reform. And no one did more to articular this view than German Chancellor Helmut Kohl. In a television interviews and speeches throughout 1990, Kohl painted a picture of a GDR transformed into "blühende Landschaften," or blooming landscapes – where people would live and work in prosperity.[30] As political leaders in both countries drafted treaties laying out the principles of a new market in the east – "private ownership, competition, free pricing," as well as the EU-styled "freedom of movement of labor, capital, goods and services" – Helmut Kohl loudly proclaimed a new *Wirtschaftswunder* for soon-to-be-ex-denizens of the GDR.[31] The Chancellor's rhetoric implicitly critiqued the GDR as desecrated and polluted. It invoked nature as a black-boxed panacea for East Germany's ills.

Kohl was not alone in his euphoria. Environmentalists also embraced the *Wende*'s promise and believed that dawning reunification offered a chance to remake Germany. Many saw the former East, and its dirty energy sector, as a *tabula rasa* for the full implementation of environmental reconstruction. Think tanks like the *Insitut für Energetik* (IfE) developed plans for the complete conversion of the former GDR to renewable energy. Rather than copy existing models of energy production, distribution, and consumption in the East, the IfE proposed a complete and radical break.[32] And it was not alone. In 1990, the *Institut für Ökologische Wirtschaftsforschung* produced a study of East German environmental pollution under the direction of scientists from East and West

30 Helmut Kohl, "Fernsehansprache von Bundeskanzler Kohl anläßlich des Inkrattretens der Währungs-, Wirtschafts- und Sozialunion," *Bulletin der Presse/ und Infomationsamt der Bundesregierung Nr. 86* (July 3, 1990). <http://www.helmut-kohl.de/index.php?msg=555> Accessed June 3, 2017. Film footage of a similar speech can be viewed at <https://www.youtube.com/watch?v=Zhxeg-1ATfg>.

31 Thomas A. Baylis, "Institutional Destruction and Reconstruction in Eastern Germany," in *After the Wall: Eastern Germany since 1989*, ed. Patricia Jo Smith (Boulder, CO: Westview Press, 1998), 109–10.

32 Sonja Boehmer-Christiansen, "Environment-Friendly Deindustrialization: Impacts of Unification on East Germany," in *Environment and Society in Eastern Europe*, eds. Andrew Tickle and Ian Welsh (Harlow, Essex: Addison Wesley Longman, 1998), 78.

Germany. Throughout the evaluation of the energy sector, agriculture, chemical production, and infrastructure, the authors repeatedly stressed the potential for a true *Wende*. "The collapse of the system and the opening of the GDR to western influence brings a number of chances and possibilities for environmentalism in both the East and the West, but also risks. An opportunity has arisen to correct the failed developments of the past."[33] At the same time, Petschow, Meyerhoff, and Thomasberger decried the siren-song of the market-system as a false "panacea"; that in fact, the history of the West German environment showed that for all the progress made, the West German political system has created just as many "environmental dead ends."[34]

Unfortunately the speed of the transition – particularly the economic collapse of the East German economic activity in 1990 – swept away any chance for deliberate ecological planning. Enthusiasm for the *Wende* created a seemingly natural momentum for neoliberal economic reform. "The rush to unification," as Konrad Jarausch argued, left the process "over reported and under-analyzed."[35] Fateful decisions like the currency reform and privatization were ratified in a matter of months, and induced the speedy wind-up of the GDR and its institutions. Administrative institutions, laws, factories and property rights were dissolved, abrogated, downsized and redistributed.[36] Most consequentially, the *Wende* induced a massive transfer of wealth from East to West, the effects of which were world-altering for former citizens of the GDR. Within three and half years of reunification over 92 percent of state firms had been privatized, with 85 percent falling into the hands of West Germans, 9 percent to international investors, and only 10 percent to East Germans.[37] Overall East German employment dropped from 9 million to 6 million by 1997 as the "East," was forced to

33 Petschow et al., *Umweltreport*, 9.

34 Petschow et al., *Umweltreport*, 11.

35 Jarausch, *German Unity*, 4–5.

36 Baylis, "Institutional Destruction and Reconstruction," 18–21.

37 Patricia Jo Smith, "The Illusory Economic Miracle: Assessing Eastern Germany's Economic Transition," in *After the Wall: Eastern Germany since 1989*, ed. Patricia Jo Smith (Boulder, CO: Westview Press, 1998), 123.

compete in the global market economy constrained by the relative strength of West German economic conditions.[38]

In the case of agriculture and energy, environmental concerns were used to diminish the value of public assets. Despite its poor reputation, East German agriculture was not on the verge of collapse in 1990. As one farm manager remarked, "we could have continued to have operated as an LPG for another five years. If we had engaged in barter or something else, it would have still worked out."[39] But sensationalized stories about pollution made East German fruits, vegetables, and meats untouchable across unified Germany. "Everybody wanted apples from Holland, the Tirol or Lake Constance. They didn't want to eat those 'poisoned' apples any longer. But then, when we talked to our colleagues from the Lake Constance, we found out that we were, according to them, not using *enough* pesticides. We sprayed ten to twelve times per year, while they sprayed eighteen to twenty times."[40] At the same time, state sponsored land reform and competition with market-oriented farmers drove a rapid collapse in the structure of East German agricultural employment. As of 1989, East German agriculture employed 850,000 full-time workers. By 1995 the number had fallen to 157,000; four years later it stood at 112,000 in 1999, an 87 percent drop.[41]

Environmental issues wrought an even faster transformation in East Germany's energy sector. As the politics of German environmentalism shifted away from forest death and sulfur dioxide to global warming and carbon dioxide, the federal government called for an immediate moratorium on brown coal mining and brown coal power plants, shuttering the industry. At the same time, they also took East Germany's four nuclear power plants offline, largely in response to recent fears raised by Chernobyl and a November 1989 accident at Greifswald, East Germany's largest plant.[42]

Even though power plants were shut down, it did not mean they lost all of their value. Shortly thereafter western energy firms moved in to buy

38 Smith, "The Illusory Economic Miracle," 131.
39 Buechler and Buechler, *Contesting Agriculture*, 104.
40 Buechler and Buechler, *Contesting Agriculture*, 109.
41 Buechler and Buechler, *Contesting Agriculture*, 178–79.
42 Dominick, "Capitalism, Communism, and Environmental Protection," 319–20.

up what remained. Despite the dirty reputation of lignite, western firms saw the brown coal reserves as potentially profitable. In 1993, the Anglo-American consortium of PowerGen and NRG Energy purchased power plants and a briquette factory in the infamous industrial south, planning to produce 20 million tons of lignite per year, while simultaneously shedding 3,000 jobs from a workforce of 9,000 people.[43] The Swedish state-owned utility Vattenfall acquired both the open-cast pit mines in Lusatia as well as the brown-coal power plant Jänschwalde in 1994. In this way, widely circulated environmental critiques of the GDR lowered the value of East German assets, creating the legal and political justification for corporate takeover.[44]

The purchase of the East German energy sector did not raise many objections from environmentalists because it occurred against the backdrop of dramatic reductions in emissions in the East. Yet this was not the result of environmental policies but instead the collapse of the East German economy beginning in 1990. According to the West German government in Bonn, carbon emissions fell from 348 million tons to 288 million tons between 1987 and 1990 in the East alone, representing an 18 percent reduction in carbon emissions.[45] German emissions continued to decline after 1990 as a result of so-called "deindustrialization." Over the course of the *Wende*, the loss of markets and import substitution drove East German companies into the red. As factories went silent, mines closed, and collective farms came apart, energy demand collapsed. Within two years of reunification, industrial production fell by two thirds, while annual energy consumption fell from 26 to 16 gigawatts in 1991. And in the 25 years since, the government estimated that deindustrialization created a 10 percentage-point drop in carbon emissions.[46] In this way the foundations of a dirty energy sector remained an integral part of Europe's energy policies, even as Germany claimed to have turned over a "green" leaf.

43 Boehmer-Christiansen, "Environment-Friendly Deindustrialization," 75.
44 Boehmer-Christiansen, "Environment-Friendly Deindustrialization," 93.
45 Boehmer-Christiansen, "Environment-Friendly Deindustrialization," 79–80.
46 Jarausch, *Rush to German Unity*, 155; Morris and Jungjohann, *Energy Democracy*, 131–32.

State Socialism's Half Life

Within a couple of years, speedy reunification and deindustrialization left many Germans with an impression of the *Wende* as leaving nothing of the GDR except *Land und Leute*.[47] Yet the GDR's 40-year history could not be banished from existence in a decade. As the federal government condemned the ecological lawlessness of the East German regime, it failed to address many of the fundamental ecological problems that had bedeviled East German planners. And so the ecological legacy of the East German state persisted, shaping economic and environmental challenges well into the third decade of reunification. By 2011, Germans found themselves debating many of the same environmental questions that had been directed towards the GDR in 1990. Brown coal mining and nuclear power, nitrate runoff and industrial livestock production, nature conservation and wild boar overpopulation, returned to the national political stage. Few Germans, however, are aware of the extent to which the GDR continues to shape these debates. Due to the *Wende*'s speed and myth making, East Germany's ecological history remains missing from the conversation.

Take for example agriculture. In the decades since the *Wende*, Germany has gone from net-importer of meat to a net-exporter, a significant milestone that challenged German politicians and economists for more than a century. Yet the rise of meat production has been accompanied by a dramatic rise in foul, unusable livestock manure – some 191 million cubic meters in 2010 alone. The manure is a major contributor to nitrate pollution in Germany's waterways. Some of it leaches out of slurry lagoons and treatment facilities after sever weather. A not insignificant amount – approximately 9.6 million liters in 2015 – spills untreated into the environment because of transportation accidents.[48]

47 Baylis, "Institutional Destruction and Reconstruction," 17.

48 Katrin Wenz and Katrin Ziebarth, "Düngerüberschüsse aus der Landwirtschaft: Gefahr für Flüsse, Seen und Meere," *Der kritische Agrarbericht* 16, no. 1 (2017): 199–200.

Where once western critics decried East Germany's Confined Animal Feedlot Operations (CAFOs) as examples of "Socialist Megalomania," extreme concentration has become the rule across reunified Germany. Between 2000 and 2013, the number of hog rearing facilities decreased from 125,000 to 25,000, an 80 percent drop. At the same time the number of pigs per facility rose from 350 to 2,163, while Germany also increased the annual number of slaughtered pigs in the country from 40 to 60 million.[49]

Livestock production, while overwhelmingly responsible for agricultural pollution, is not the sole contributor to this problem. German officials estimate that 37 percent of total nitrogen run off comes from the over-application of agrochemicals to fields. This is partially a product of the massive size of eastern Germany's fields, created over the course of the 1960s and 1970s. These fields, some as large as 8 square kilometers or almost 2,000 acres, are particularly susceptible to runoff from precipitation. While the type of crops grown in the former east have changed, the scale of production has not. Rape seed/canola, corn, and wheat are cultivated across the five new *Bundesländer* and make up 27 percent of all arable land in Germany.[50] As of 2015, 35 percent of Germany's 2.6 million ha of maize were cultivated for biofuels, despite ongoing questions about the sustainability of the practice.[51] The dangers of agricultural pollution have only grown since the collapse of the GDR. Nitrate and phosphorous pollution continue to overwhelm Germany's largest coastal bodies of water, the North Sea and Baltic Sea, with massive dead zones caused by eutrophication.[52]

A similar red line runs right through the history of the energy sector since 1989, made visible by the *Energiewende*. The immediate suspension of nuclear power in 2011 had the ironic effect of bringing the brown coal

49 Jutta Sundermann, Leonie Dorn, and Reinhild Benning, "Wachsen oder Trinken? Grenzen der Konzentration in der Tierhaltung und der Streit um die Düngeverordnung," *Der kritische Agrarbericht* 15 (2016), 202.

50 David Crossland, "Climate Change's Clear Winners: Europe's Wild Boar Population," in *Der Spiegel Online International*, November 25, 2009, accessed June 16, 2014, <http://www.spiegel.de/international/zeitgeist/climate-change-s-clear-winners-europe-s-wild-boar-population-exploding-a-663411.html>.

51 Wenz and Ziebarth, "Düngerüberschüsse," 203, n. 13.

52 Wenz and Ziebarth, "Düngerüberschüsse," 199–200.

industry back to life, due to a sizeable shortfall in energy supply. Renewables, while producing at higher levels than anywhere else in the world, still could not cover the entire energy demands of the country, and the brown coal industry raced to fill the shortfall. In 2013, Germany produced more electricity from brown coal than in any year since 1990.[53] Although it fell off this peak in the succeeding years, brown coal remains Germany's second largest fuel.[54]

The resurgence of brown coal revitalized old debates that dated back to the last days of the GDR. One major area of concern was mine expansion. During the late 1980s, westerners criticized the Lusatia operation for driving people from their homes. Opencast mining is a comparatively cheap technique, since it involves digging long trenches at the surface level, as opposed to tunneling deep into mountain sides. Unfortunately since brown coal is found closer to the surface than other forms of mining, it requires immense amounts of land. The Jänschwalde mine, for example, presently occupies 23 square miles.[55] Between 1960 and 1980, the Lusatia mines swallowed some 70 villages in search of more lignite, a point critics used to decry the monstrous nature of the communist state.[56] After the *Energiewende*, however, the lignite behemoth staggered back to its feet, this time in service of a capitalist state hungry for electricity.[57]

And like 1989, a protest movement emerged to oppose the industry. Near Vattenfall's Lusatia holdings, the village of Atterwasch received national and international attention in 2014 after residents and environmentalists

53 Stefan Wagstyl, "German Coal Use at Highest Level Since 1990," *The Financial Times*, January 7, 2014, accessed June 15, 2017, <https://www.ft.com/content/e6470600-77bf-11e3-807e-00144feabdc0?mhq5j=e1>.

54 "Bruttostromerzeugung 2016," Destatis, *Das Statistische Bundesamt*, <https://www.destatis.de/DE/ZahlenFakten/Wirtschaftsbereiche/Energie/Energie.html>.

55 <https://de.wikipedia.org/wiki/Tagebau_J%C3%A4nschwalde>, accessed June 20, 2017. Notably, this is after renewables.

56 DeBardeleben, "'The Future Has Already Begun'," 148.

57 Andrew Curry, "Germany Plans to Raze Towns for Brown Coal and Cheap Energy," *National Geographic Magazine*, February 13, 2014. <http://news.nationalgeographic.com/news/energy/2014/02/140211-germany-plans-to-raze-towns-for-brown-coal/>, accessed June 15, 2017.

protested plans to destroy their homes.[58] Vattenfall's boosters countered with familiar arguments: the company was the region's only remaining, major employer; it kept 25,000 Germans living in the region; its corporate responsibilities also included funding local culture like the symphony as well as the regional ice hockey and soccer teams.[59] In essence Vattenfall was performing the same functions that large socialist operations, like collective farms and factories, had once provided under state socialism.[60]

Pollution, however, was not East Germany's only lasting legacy. Ironically reunified Germany's "green" reputation was in many ways built on top of a conservation footprint laid out by the East German regime. This was especially true in Lusatia. Beginning in the 1980s, the East German regime implemented pilot restoration programs for its lignite mines, filing in old pits to cultivate them with maize, wheat, and oats.[61] After reunification, the Federal Government poured even more money into the remediation of Lusatia. They established the Lusatian and Middle German Mining Administrative Company (LMBV) to oversee the creation of what would become known as the *Lausitzer Seeland*, or

58 Melissa Edy, "German Village Resists Plans to Strip It Away for the Coal Underneath," *The New York Times*, February 18, 2014, <https://www.nytimes.com/2014/02/19/world/europe/german-village-resists-plans-to-strip-it-away-for-the-coal-underneath.html>, accessed June 16, 2017.

59 Sven Becker, "Vattenfall Overpowers Coal Critics," *Der Spiegel Online*, October 13, 2013, <http://www.spiegel.de/international/germany/energy-giant-vattenfall-presses-on-with-coal-mine-expansion-plans-a-930828.html>, accessed June 17, 2017.

60 Anke Blumenthal, "Braunkohle-Tagebau Jänschwalde wird nicht erweitert," *Rundfunk Berlin Brandenburg/24*, March 30, 2017, accessed June 18, 2017, <https://www.rbb-online.de/wirtschaft/thema/braunkohle/beitraege/brandenburg-lausitz-braunkohle-tagebau-jaenschwalde-wird-nicht-erweitert.html>. Yet unlike state socialism, market forces have had a much more direct effect on the company's future. And in 2017, the Swedish government pulled the plug on Vattenfall's German operations. Recognizing the continued growth of renewables in Europe as well as their own commitment to combating global warming, they sold Vattenfall's Lusatian holdings to a Czech company, EPH, which put expansion plans on hold, sparing villages like Atterwasch – at least until sunnier economic skies return.

61 *Neue Deutsche Bauernzeitung*.

Lusatian Lake District in 1994.[62] Today the migratory birds, rare wild flowers, and "old growth" hardwood forests that ring its lakes belie the park's industrial past. Ecological reconstruction required the flooding of 24 open pit mines creating artificial lakes that covered 54 square miles. The LMBV treated its water for heavy metals and other pollutants while it planted trees to reduce erosion, and introduced aquatic species of fish, amphibians, and plants.[63]

More broadly Lusatia represented one of thousands of such programs implemented under the broad umbrella of *ökologischer Umbau*, or ecological reconstruction. Following the German Chancellor's edict of "blooming landscapes," these programs argued that environmentalism could offset unemployment while also improving public health. In 1991 the "National Solidarity Action for Ecological Restructuring" program called for the investment of 5 billion DM in similar projects. And by 1992 the Ministry of the Environment had identified 12,000 contamination sites, completed work on 600 environmental projects, while also employing 50,000 people. The vast majority of these jobs went to the *Sanierung*, or restoration of industrial disaster zones, like Lusatia.

Yet even ecological restoration had an East German precedent. In the waning days of the GDR's existence, the East German council of ministers created numerous national parks, setting aside four percent of the country's total area for conservation, including five national parks, six "biosphere" reserves, and three nature parks.[64] This was not, however, merely a last-ditch swerve towards environmentalism but rather a political evolution that followed from East Germany's existing form of environmental thought. Contrary to its reputation, the East German regime had promoted forms of nature protection around the principles of mixed-usage, the enjoyment of nature, and managed economic use.[65]

62 Peter Mellgrand, "Life after Lignite: How Lusatia Has Returned to Nature," *The Guardian*, September 10, 2014, accessed June 20, 2017, <https://www.theguardian.com/environment/2014/sep/10/lusatia-lignite-mining-germany-lake-district>.

63 Mellgrand, "Life after Lignite."

64 Uekötter, *Greenest Nation* 138.

65 See Moranda, *People's Own Landscape*.

During the *Wende*, eastern Germany's ecological reconstruction, while based in western-style conservation, was implemented on top of East Germany's existing system. The so-called European Green Belt, a transnational conservation program which is seeking to establish a contiguous corridor of migration in the footprint of the Iron Curtain, from the Baltic to the Black Sea, is the most prominent of these new programs.[66] But these programs belong alongside reconstruction programs like the Lusitian Lake District, since they make recreation, natural habitat restoration, and animal species protection primary goals. And it was against the backdrop of ecological reconstruction that the wild boar problem must be understood.

While the spike in the boar population appears new, it too was a problem that dated back to the GDR. It was East Germany's mixed usage "ecological regime" that contributed to the first major spike in the wild boar population in the 1970s. The principles of industrial meat production placed some 11–13 million pigs into the lands of East Germany. Coupled with the broader adoption of industrial confinement around Europe, the German continent had more pigs living on it by 1989 than ever before. At the same time, the GDR witnessed a remarkable reforestation between 1971 and 1989, as the creation of larger fields also produced more marginal plots of marshes and forests.[67]

East German Wild life managers noted the link between growing pig populations and increased forest cover. Borrowing from Aldo Leopold, they studied the creation of edge spaces around the country. It was during this time they measured a remarkable spike in the country's population, from roughly 20,000 wild boars in 1963 to more than 160,000 by 1989.[68] This so-called ungulate irruption laid the first baseline for post-reunification ecologists.

66 Caroline Fraser, *Rewilding the World: Dispatches from the Conservation Revolution* (New York: Metropolitan Books, 2009), 79–81.

67 Thomas Fleischman, "'A Plague of Wild Boars': A New History of Pigs and People in Late 20th Century Europe," *Antipode: A Radical Journal of Geography* 49, no. 4, 25.

68 Fleischman, "'A Plague of Wild Boars'," 25.

Bearing this in mind, the European Green Belt and the broader program of *ökologischer Umbau* may have served to augment a process that was already underway. The fundamental premise of the Green Belt is to promote the free movement of wild life along the corridor, while also connecting them to preserves once isolated from one another. Thus "new" parks like the Lusatian Lake District become integrated into a growing network of conservation zones. The GDR's post-socialist landscape can serve as both an epicenter and conduit for expanding populations of wild boars, both radioactive and "natural." Surely enough hunting clubs in the Lusatian Lake District have reported a spike in wild boar sightings, even as the park offers various cuts of wild boar meat for sale on its website.[69]

Conclusion

The transition to a sustainable energy economy is still an open question. Despite the massive progress the German government has made, many challenges remain. As one journalist noted of the road ahead, "The *Energiewende* will come, but it will be more difficult, more expensive, and require more time than many thought during the *Wende* euphoria, even with the visage of Fukushima on our minds."[70] As we have seen, the debates about the future of the *Energiewende* are deeply embedded in the historical myth of the last *Wende*. Yet the environmental history of post-reunification Germany is bigger than just the question of energy. Food production, land

69 Ulrike Elsner, "Wildschweinbestände im Spree-Neiße-Kreis wachsen dramatisch," *Lausitzer Rundschau*, April 11, 2017, accessed June 25, 2017, <http://www.lr-online.de/regionen/cottbus/Wildschweinbestaende-im-Spree-Neisse-Kreis-wachsen-dramatisch;art1049,5935972>.

70 Wolfgang Becker, "BDS-Energiesymposium: Die Wende von der Wende von der Wende," *Bund der Selbtändigen Baden-Württemberg e. V.*, September 20, 2012, accessed June 10, 2017, <http://www.bds-gewerbevereine.de/v2/public/land/content.aspx?id=63f29542-17dd-4ed1-a813-8b1954cc2dc3>.

usage, and industrial development will continue to be debated for years. Those debates, however, cannot happen without a full-reckoning with the East German's ecological history. Its forgetting over the course of the 1990s shaped the contours of environmental questions to the present. Yet when we consider the case of unruly, radioactive boars, we are reminded of East Germany's on-going legacy, shaping the lives of people and animals, like decaying isotopes of a socialist utopia, combusted long ago.

CHRISTINA SCHWENKEL

Shrinking Green Cities: Trees and the Afterlife of Eco-Socialist Planning in Vietnam

In a recent article that questions the uniqueness of the so-called socialist city, Sonia Hirt argues that one of the lasting legacies of socialist urban development is, ironically, its greenness; its verdant public parks and grassy open spaces. I say ironic because in conventional thinking, socialist cities are often associated with unlivable, concrete settlements[1] – the infamous asphalt jungles reified in films such as *This Ain't California* (2012) – rather than lush urban environments. And yet, Sonia Hirt shows in her comparison of two European cities with intersecting imperial histories, Budapest and Vienna, that public green space in the former exceeded that of the latter by nearly four times.[2] A reading of manuals and journals on urban planning in the early years of the German Democratic Republic (GDR), such as *Deutsche Architektur*, shows an explicit concern with ecological planning in debates over the *Wohnungsfrage*, or housing question, and the (re)construction of planned socialist cities, such as Hoyerswerda, Cottbus, and Rostock. Moreover, Ruth May has shown how early drafts of the master plan for Stalinstadt, as the first socialist new town, contained "spaciously planned connected green zones" with recreational amenities such as lakes and parks as an expression of the GDR's Sixteen Principles of Urban Development of 1950 to help residents realize the socialist good life.[3] Even with the

1 Heinrich Niemann, "Was in einem Reiseführer stehen sollte," in *(Keine) Platten Geschichten: Marzahn erzählt*, ed. Katrin Rohnstock (Berlin: Ronhstock Biografien, 2004), 159–164, 160.

2 Sonia Hirt, "Whatever Happened to the (Post)socialist City?" *Cities* 32 (2013): 29–38, 32.

3 Ruth May, "Planned City Stalinstadt: A Manifesto of the Early German Democratic Republic," *Planning Perspectives* 18, no. 1 (2003): 47–78, 65.

shift to mass-produced housing as part of the *Wohnungsbauprogramm* of the 1970s there remained an emphasis on *Grünanlagen*, as Eli Rubin has also demonstrated in his social history of Marzahn, a massive housing settlement on the eastern edge of Berlin.[4] These works remind us that the brighter future of socialism envisioned by these planners was one that was unambiguously green.

And yet, to this day, socialism remains characteristically associated with drab greyness.[5] Images of inhumane, barren landscapes devoid of color and vibrancy signify then a pathological condition of dull life without nature and its possibilities for renewal. In a Cold War triumphalist narrative, capitalism, on the contrary, marks an end to the dreariness, with its vibrant colors, suggestive of free choice and greener futures, bursting onto the scene to save despairing people who long for the connection to, and curative powers of urban nature. In this chapter, I inverse this paradigm and argue for the opposite, while pointing to a paradox: how the contemporary era of green – green cities, green buildings, green architecture and so on – with its promises of ecological sustainability in a moment of grave environmental uncertainty has meant the death of nature in the socialist city. In other words, it is under capitalist forms of accumulation, property relations, and social organization that the greenery of trees and the grassy lands of socialist cities have given way to the drabness of pavement and concrete. As Walter Benjamin once claimed: "it is *fallen* nature which bears the imprint of the progression of history" (emphasis added) and the ruined landscapes it leaves in its wake.[6]

At the heart of post-socialist urban redevelopment has been the dismantling of socialist ecological planning. Subjected to speculative value, public green space has been privatized and appropriated for commercial use and other purposes. As Hirt has observed of Sofia, public

4 See Eli Rubin, *Amnesiopolis: Modernity, Space, and Memory in East Germany* (New York: Oxford University Press, 2016).

5 See Krisztina Fehérváry, *Politics in Color and Concrete Socialist Materialities and the Middle Class in Hungary* (Bloomington: Indiana University Press, 2013).

6 Walter Benjamin, *Origins of the German Tragic Drama*, trans. John Osborne (New York: Verso, 1998), 177.

green space is rapidly shrinking: an estimated 30 percent has been converted to private use since the collapse of communism in Bulgaria.[7] In what follows, I provide another example of the undoing of what I refer to elsewhere as "eco-socialist planning" in the industrial town of Vinh, Vietnam, rebuilt as a "model" socialist city after the end of the air war in 1973 with the technical assistance of East Germany.[8] During the US war in Vietnam, American air power annihilated this industrial port city in north central Vietnam, positioned at the start of the so-called Hồ Chí Minh trail. As a prestige project, Vinh's reconstruction presented a unique opportunity for the East German government to bolster its international public image and its claims to moral superiority over the FRG. The small provincial city would thus become a strategic site in this Cold War rivalry. In this chapter I examine how the global Cold War played out ideologically in a decolonizing country through planning practices that centered on green city making as a means to national recovery that would hasten the expansion of global socialism while showcasing East Germany's technological ascendency.

Green technologies traveled to Vietnam through circuits of "fraternal solidarity" that informed the ideological practice of socialist internationalism at the time. In what follows, I show how new forms of capitalist restructuring since economic reforms signal the slow disassembling of eco-socialist planning with its rational, green design of mass housing that aimed to produce more healthful and just living environments for postwar populations. The material and ideological foundation to this modernist planning coalesced through an attention to the role and place of nature in the city to produce new urban socialist persons. As Rubin has shown, the biophysical environment proved to be an effective medium through which the state created socialist subjectivities in the GDR.[9] In Vietnam, people came to identify closely with, and appreciate, their eco-housing and

7 Hirt, "(Post)socialist City," 32.
8 See Christina Schwenkel, "Eco-Socialism and Green City Making in Postwar Vietnam," in *Places of Nature in Ecologies of Urbanism*, eds. Anne M. Rademacher and K. Sivaramakrishnan (Hong Kong: Hong Kong University Press, 2017), 45–66.
9 Rubin, *Amnesiopolis*, 100.

communal green spaces, and the threat of their loss has inspired a revalorization of their obsolescent complex and the foreign designers responsible for its creation.

In architecture and planning circles around the world today one finds a fascination, if not fetishization, with all that is deemed *green*. Along with its "eco" complement, "green" has become a new branding device and measure of value. This is also the case in Vietnam, where "greenness" is thought to herald a new era. In the real estate sector, objects, sites, and property predicated with its prefix acquire a higher market value: green housing, green space, green architecture, and green cities. Such concepts mark a new, not entirely democratic regime of sustainable things that generate new forms of class distinction through the ability to consume that which is deemed ecologically viable. This consumption then marks an emerging environmental sensibility and the production of eco-responsible citizens with developed "green" tastes. Claims to the green industry as a novelty (and a public good) should not go undisputed, however; greenness in all its connotations needs to be historicized, contextualized and contested to reveal how it has enabled stratification and redrawn lines of exclusion. My aim then is to point to an earlier iteration of urban ecology during the global Cold War that was not a response to neoliberal capitalism and environmental crisis as we see today, but – at least in the case of Vietnam – preceded it. I propose to do so by looking at the intersection between socialist architecture and urban ecology; more specifically, the relationships between mass housing and trees that were considered vital to enabling a more healthful and egalitarian society for workers and their families. As Rubin has argued of Marzahn, trees served as an important metaphor for the futurity of socialism, forming a topography "of space, time, and memory" that was set into motion by state urban policy.[10] This was similar to the case in Vinh, but in this context of decolonization and "fraternal solidarity," trees also celebrated the renewal of the destroyed city and its global connectivity to the rest of the communist world in an optimistic moment of socialist nation building.

10 Rubin, *Amnesiopolis*, 122.

Germinating Socialism: The Meaning of Trees in Vietnam

Around the world, trees are charismatic objects imbued with symbolic meaning that shape identity and define cultural engagements with the biophysical environment. Trees, anthropologists argue, have social lives that serve different social, political and ecological purposes across time and space, and provoke different affective responses.[11] There is an important duality to trees noted in the literature: they both index the past and allude to the future, embodying memories of what has come to pass on the one hand, while inciting hope for what might be on the other. While trees signify life, vitality, and regeneration through their bountiful growth, they can also connote decay and destruction with their slow aging and eventual death.

In Vietnam, trees are hopeful symbols of future productivity that are also explicitly nationalist. The Vietnamese nation is often compared with bamboo, for instance, insofar as both the people and the plant are known to be resourceful and resilient. Reforestation projects – including the planting of "peace trees" by US veteran returnees – have stood as signifiers of postwar recovery (and renewed international relations). In cities, urban trees, especially those with deep roots in rural histories, are likewise suggestive of the promise of future prosperity.[12] Mature trees are often worshipped on account of the spirits that inhabit them – spirits that assure material abundance in the present if cared for adequately. For this reason, one can often find incense and small altars placed among the twisted roots of ancient banyan trees in the capital city of Hanoi. As Le Trong Cuc has observed, the veneration of tree deities, including in old ficus trees with large canopies, was recognized as helping both people and country (*hộ quốc hộ dân*).[13] Given the complex geopolitical, cultural and economic attributes that people ascribe to them,

11 See Laura Rival, ed., *The Social Life of Trees: Anthropological Perspectives on Tree Symbolism* (Oxford: Berg, 1998).

12 Christina Schwenkel, *The American War in Contemporary Vietnam: Transnational Remembrance and Representation* (Bloomington: Indiana University Press, 2009), 39–42.

13 Le Trong Cuc, "Vietnam: Traditional Cultural Concepts of Human Relations with the Natural Environment," *Asian Geographer* 18, no. 1–2 (1999): 67–74, 72.

what perhaps can trees and their corresponding green spaces tell us about the socialist city at the height of the Cold War?

In Vietnam, trees have long been associated with postcolonial nation building. This is illustrated in stock images of Hồ Chí Minh, who was typically shown either communing with nature or with children, both of which were sites for the germination of socialism, which would take deep root in society and offer its bounty to future generations. The most well known example was the annual ritual planting of a lunar new year (Tết) seedling carried out by Hồ Chí Minh and his famous maxim: *Vì lợi ích mười năm thì phải trồng cây, Vì lợi ích trăm năm thì phải trồng người*, which can be translated as "for the benefit of a decade, one should cultivate trees, for the benefit of a century one should cultivate people." Linguistically, the slogan plays on the Vietnamese verb *trồng*, which equates the physical and emotional labor of planting with the nurturing human and nonhuman things.

Internationally, trees have served as important geopolitical devices and valuable objects of diplomatic gifting. During his visit to New Delhi in February 1958, Hồ Chí Minh planted a Vietnamese đãi tree at the *Samadhi Rajghat* Mahatma Gandhi memorial, while President Prasad, in return, presented Hồ Chí Minh with a Bodhi tree, the seed of which reportedly came from the sacred fig tree under which the Buddha attained nirvana. Such exchanges showed the central role that nationalized nature – each tree representing its nation's ecological history – played in the cementing of international support for the revolutionary government in the years leading up to the war with the United States, when American air power would sow its seeds of destruction.

Trees, also in urban settings, played an important role in the revolution and in the reification of utopian visions of longevity ("Xã Hội Chủ Nghĩa Muôn Năm" or socialism forever). Trees produced the fruits of the nation (literally and metaphorically), as well as loyal citizens framed as happy healthy families poised between a traditional rural past and a modern urban future, without forsaking one for the other. Trees thus serve as a temporal and biophysical conduit between these interdependent worlds and economies, bridging nature with technology. Such ideas are commonly communicated by means of urban propaganda posters, often through women's bodies and especially female ethnic minorities who are marked as "closer" to nature. In one sign, a highland Tai woman holds a green sapling in her open hand, signifying future abundance, while clasping a book on science and technology close to her chest. The effect is to merge city and nature, *techne* and *metis*, lowlands and highlands, into a unified polity.

That trees and urban greenery served to define the city became apparent to me during my fieldwork on the bombardment of Vinh in 2010–11. During this time, residents described to me the annihilation of place not only in terms of the presence of rubble, but also through the absence of nature. In other words, there was an emphasis on the material devastation of both the biophysical and the built environment. Green destruction was of course a strategic goal of the US military. During the war, trees had served an important protective function; it was under the forested canopy that combatants mobilized and sought cover from the enemy. The use of chemical defoliants put an end to such invisibility. Cities, too, suffered from the annihilation of its greenery. In the recollection of one Vietnamese man: "The city was flattened [by bombs]. I could look out to the horizon without any tree blocking my view."[14] This is the first phase of Benjamin's fallen nature, where ecology falls victim to the violence of imperialism and ushers in a new life cycle of the city: from urban death to rebirth, material and ecological regeneration through the collective labor of reconstruction, including the sowing of new seeds.

Socialist Trees: Sowing New Seeds through Ecological Planning

Trees were also central to rebuilding the postwar city and to transmitting socialist ideology. This occurred both biophysically in the plant itself, and metaphorically in its symbolic meaning of new life.[15] During the Cold War,

14 Personal interview with author, Vinh City, May 28, 2011.

15 For more details on the eight-year project that involved more than 200 East German technicians, see Christina Schwenkel, "Spectacular Infrastructure and its Breakdown in Socialist Vietnam," *American Ethnologist* 42, no. 3 (August 2015): 520–534.

the transfer of socialist planning expertise gave rise to a range of experimental urban forms and techniques of design across the decolonizing world. GDR concepts of planning traveled widely overseas, including to North Korea, Tanzania, and Vietnam, where urban ecology in a tropical climate, and the planting of new trees that would transform society, played a pivotal role in the master plan for rebuilding a devastated city and re-housing its evacuated population. The seven-year project in Vietnam was unique because experts lived on site for an average of one to two years (and some even longer), which allowed them a deeper grasp of the ways in which space was actively lived, experienced, produced, and made meaningful. In collaboration with local experts,[16] they translated their sense of lived space into the conceived, abstract space of planning documents, maps and blueprints.[17] Their planning objective was to rationally manage urban space and impose order on what they perceived to be disorderly landscapes as a result of spontaneous settlements that formed as residents slowly returned to their cratered city and built random makeshift houses on plots of empty land. In so doing, they drew on conceptions of *die grüne Stadt* in their designs that were circulating in East Germany at the time,[18] supported by scientific calculations of how to best attain a balance between traffic, buildings, and green spaces to improve the lives of the laboring class, and to maximize industrial productivity. For example, the central boulevard that cut across the city was originally designed as "Palm Alley" (see Figure 4). Like in East Germany, the separation of traffic and pedestrians provided generous public spaces for residents to move about unhindered. In Vietnam, Palm Alley offered residents a broad, tree-lined street for families to stroll along, one that lit up at night with newly installed street lights. The plan

16 Some of these experts had trained overseas, also in East Germany. On the ecological training of Vietnamese students in the Soviet Union and East Germany, see Christina Schwenkel, "Socialist Mobilities: Crossing New Terrains in Vietnamese Migration Histories," *Central and Eastern European Migration Review* 4, no. 1 (June 2015): 13–25 and Schwenkel, "Eco-Socialism," 50.

17 See Henri Lefebvre, *The Production of Space*, trans. Donald Nicholson-Smith (Oxford: Blackwell, 1991).

18 See Hans Schmidt, Rolf Linke, and Gerd Wessel, *Gestaltung und Umgestaltung der Stadt: Beiträge zum sozialistischen Städtebau* (Berlin: Verlag für Bauwesen, 1969).

thus merged the technical – electric infrastructure – with the "natural" – a canopy of native trees – to suggest social stability (an end to war) and a more equitable postwar city to be experienced by all. Here urban nature became a means of social inclusion, in contrast to the exclusions of colonial urbanism prior to the revolution.

Figure 4: Palm Alley as part of the GDR's master plan for the reconstruction of Vinh. Courtesy of the German Federal Archives.

Socialist urban planning was an exact science in Vietnam as it was in East Germany.[19] Through statistical modeling, planners drafted a comprehensive master plan that closely resembled the functional city of the Athens Charter put forth by Le Corbusier and the International Congress of Modern Architecture in 1943.[20] Vietnamese architects described the plan to me, more than 30 years later, as resembling that of an open hand: education and housing were concentrated in the palm (the city center), with each finger, or

19 Elke Beyer, "Planning for Mobility: Designing City Centers and New Towns in the USSR and the GDR in the 1960s," in *The Socialist Car: Automobility in the Eastern Bloc*, ed. Louis Siegelbaum (Ithaca, NY: Cornell University Press, 2011).

20 Louis Mumford, *The CIAM Discourse on Urbanism, 1928–1960* (Boston, MA: MIT Press, 2002).

main transportation artery, leading to a sector of industry.[21] Green belts, as public spaces for recreation and designated zones for urban agriculture, served as a bridge between work and areas of dwelling. The model of the functional city had long been applied (and modified) to the spatial organization of other socialist cities in Europe, and was also influential in planning circles in East Germany.[22] It fit neatly within the parameters of the GDR's Principles of Urban Development from 1950, which called for a harmonious balance between work, home, culture, and leisure. As Brigitte Le Normand has argued of Belgrade, functionalist socialist planning, with its attention to the collective use of green space (rather than private gardens), extended the benefits of modern living to all urban residents.[23] Urban nature – recognized as pivotal to the health and restoration of the labor force as far back as early Soviet planning – would augment worker efficiency through routine fitness activities.[24] This was appealing to the Vietnamese planners I interviewed who worked on the reconstruction project; they valued the rational ordering of space and the attention to harmony in ecological planning as integral to industrialization. For them, urban order as modern and desirable. At the same time, they were critical of the emphasis on leisure and its associations with consumption that spoke more to the higher standards of living in East Germany than to the pragmatic possibilities for postwar recovery and the formation of a distinct "Vietnamese" socialist modernity.

Trees and green planning also played a critical role in the construction of sustainable mass housing. Scholarship on the reconstruction of destroyed cities tends to prioritize the built form, with less attention paid to the regeneration of nature. One can see this, for example, in literature on postwar Germany that focuses on *material* rehabilitation rather than the biophysical,

21 Christina Schwenkel, "Socialist Palimpsests in Urban Vietnam," *ABE Journal*: Architecture Beyond Europe 6 (2014): 9.

22 See Bruno Flierl, *Gebaute DDR: Über Stadtplaner, Architekten und die Macht. Kritische Reflexionen 1990–1997* (Berlin: Verlag für Bauwesen, 1998).

23 Brigitte Le Normand, *Designing Tito's Capital: Urban Planning, Modernism and Socialism in Belgrade* (Pittsburgh, PA: University of Pittsburg Press, 2014), 96.

24 See Stephen Bittner, "Green Cities and Orderly Streets: Space and Culture in Moscow, 1928–33," *Journal of Urban History* 25, no. 1 (1998): 22–56.

with narratives of postwar progress privileging the appearance of new buildings, not trees.[25] Literature on the construction of new towns, by contrast, takes the relationship between urban nature and the built environment more seriously.[26] In Vinh, we see similar ideas about the rational planning of urban ecosystems in the master plan for a *Wohngebiet*, or housing estate, the first of its scale in Vietnam. This sprawling assemblage of five-story apartment blocks in the city center – not on its margins like in the new towns – between industrial areas in the north and south of the city would allow for a quick and convenient commute to the workplace by bicycle or on foot, leaving more time for recreation and other activities after working hours. Intended to house 15,000 workers and civil servants, the apartments offered private indoor amenities – a luxury for residents who had lived for years in communal housing – and convenient public facilities and services within walking distance. In keeping with Corbusian ideas of scale at the time, vertical living, rather than single-family houses, would translate into more available land for communal green spaces. According to the conceptual design for the housing estate in Vinh, parks and greenery were to constitute 7 percent of the entire 34-hectare complex, averaging 1.5 square meters per person.[27] Up to 50 percent of these open spaces were to provide arboreal cover.[28]

Blueprints from 1974 show how rebuilding and re-greening the urban landscape were inextricably linked. Over the course of seven years, architects from both countries worked alongside landscape engineers flown in from East Germany for shorter missions to scrupulously design the grounds of the *Wohngebiete* to make them more sustainable and people-friendly.

25 See David Crew, *Bodies and Ruins: Imagining the Bombing of Germany, 1945 to the Present* (Ann Arbor: University of Michigan Press, 2017).

26 See in particular Katherine Lebow, *Unfinished Utopia: Nowa Huta, Stalinism, and Polish Society, 1949–1956* (Ithaca, NY: Cornell University Press, 2013) and Rubin, *Amnesiopolis*.

27 Bundesarchiv (BArch) DH1 28549, "Konzeption für die Bebauung Quang Trung," May 15, 1974.

28 Udo Purtak, *Voraussetzungen und Entwicklung des Wohnungsbaus unter humiden tropischen Bedingungen dargestellt am Beispiel des Aufbaus der Stadt Vinh in der Sozialistischen Republik Vietnam* (PhD dissertation, Technical University of Dresden, 1982), 148.

The result was an "open landscape design" or *Freiflächengestaltung* that responded to Vietnamese "climate-related living habits" (*klimabedingte Lebensgewohnheiten*).[29] These designs, drawn up collaboratively in Vinh, were based on the sociological study of cultural practices; namely through in situ observations while the experts were embedded in "the field."

Though they deployed similar methods of social science to generate seemingly objective data as colonial urbanists did,[30] they did not do so as a strategy of rule nor with the intent to subjugate and profit from the Vietnamese orientalist assessments of Vietnamese lifestyles, especially when it came to claims that they lived "closer to nature," a classic anthropological trope of the native Other.

The application of these socialist forms of intimate knowledge was framed by both sides as liberatory, pioneering, and transformative.[31] The housing blocks were spaced at a calculated distance from one another to allow for a greenbelt design that followed a cyclical pattern of housing block→access road→row of trees/green space→building, and so on.[32] The large open spaces between the buildings enabled substantial wind flow through the complex (but not too much to negatively impact the young seedlings), and generous public areas for parks and playgrounds. Shade from the tropical sun was likewise a priority that motivated the replanting of the city. Re-cultivation with diverse native plants and trees that encircled the housing blocks and lined the pathways and perimeter of the complex was to contribute to a reduction in solar heat and humidity in the housing blocks, while offering shaded communal areas for social and political activities (Figure 5). In the winter, the dense shrubbery would also protect the residents from strong, cool winds.[33] These planned green spaces, including a refurbished stadium and a central park fountain (that was rarely running

29 Purtak, *Aufbaus der Stadt Vinh*, 145.

30 See Bernard S. Cohn, *Colonialism and Its Forms of Knowledge: The British in India*. (Oxford: Oxford University Press, 1996).

31 The model they produced was celebrated across Vietnam and featured in architectural journals and government conferences. The older generation of architects in Hanoi still sees the collaborative design as *the* model for planning.

32 Schwenkel, "Eco-Socialism," 58.

33 Purtak, *Aufbau der Stadt Vinh*, 147.

due to shortages), were vital to creating a sense of normalcy in the postwar years by offering rural migrant workers a new urban future that differed radically from their previous lifestyles and forms of dwelling.

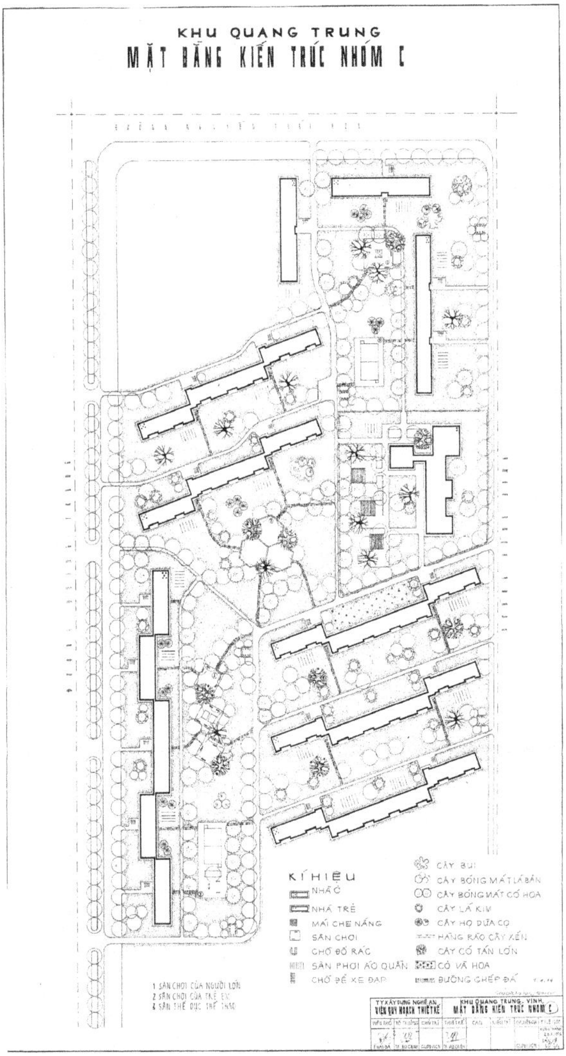

Figure 5: Area C of the housing estate. Trees are represented by circles. Courtesy of the German Federal Archives.

As I have argued elsewhere, the unique eco-socialist design of the complex brought the city widespread recognition for its resource efficiency and environmental rationality.[34] Because there was no intact infrastructure to support the rebuilding of the city – almost all machinery and equipment had to be imported – planners were forced to work with and harness the forces of nature in their planning of the *Wohngebiete*. In so doing, climate-oriented construction (*klimagerechtes Bauen*) emerged as an environmentally pragmatic solution to the technical problem of a lack of material resources. It also made more ecological sense for a humid tropical environment. As a starting point, GDR technicians and planners carried out an extensive analysis of Vinh's tropical climate, focusing on, for instance, wind frequency and directionality along with correlations between human comfort zones and levels of atmospheric humidity.[35] Based on colonial-era statistics that went back to 1907 when the French empire collected data in an effort to make the Vietnamese landscape more knowable and governable they made projected calculations about the climate's impact on the built environment before they could construct what today would be considered a rudimentary form of "green" building.[36] Based on these data, the planners arranged the blocks spatially in a zigzag pattern at an angle of 23 degrees to facilitate maximum air circulation through the buildings (see Figure 5). The northwest orientation of the blocks with a horizontal row design and an external corridor encouraged *Querlüftung*, or cross-ventilation as well as natural lighting to reduce demand for already scarce electricity.[37] Such design was considered to be hygienic and healthful: good ventilation in open, airy houses reduced illness and exposure to damp and moldy conditions, especially in the cold wet winters.

In interviews, Vietnamese planners commended the "science of wind" that informed the design of the complex, how it channeled air flows and sunlight to produce better living environments, not unlike Vietnamese geomantic practices (feng shui), or phong thủy, they explained. In the words of one retired construction engineer who still lives in the blocks: "The complex

34 Schwenkel, "Eco-socialism," 57.
35 Purtak, *Aufbau der Stadt Vinh*, 15–18.
36 Compare with Cohn, *British in India*.
37 Schwenkel, "Eco-socialism," 58.

was well designed; it had good feng shui. There were open spaces. Every door invited wind; each unit was cool in the summer. That house, this one – all were equally ventilated. This wasn't like Hanoi, where air flows were blocked."[38] Vietnamese experts extolled how the master plan had intentionally limited the height of the buildings, unlike neighboring high-rises today, which block the movement of air through the city – a point to which I return below. The residents in the housing blocks, where I also lived during my fieldwork, tended to agree. Some interpreted the staggering of the blocks as an attempt by East German engineers to apply phong thủy to the design of the complex, which they found to be quite fruitful. (In interviews, the East German experts denied any attempt to be geomancers, though they were aware of this interpretation and the overlap between "rational" and "superstitious" approaches to planning.) Others understood its spatial organization as a material gesture toward solidarity. If flying overhead, many people told me, one can see the blocks arranged in the shape of "VĐ" (Việt Đức or Vietnam-Germany), an acronym that is also imprinted over the entrance to each building. These types of myths that circulated reveal the ways in which urban residents, most of whom were rural migrants, drew on cultural frameworks to interpret and make sense of *high-rise* mass housing, which at the time was a culturally unfamiliar spatial form and style of dwelling.

That urban ecology played a pivotal role in the scientific thinking around the master plan for rebuilding a devastated city reveals how eco-socialist techniques were essential to the high modernist project of building socialism. With its climate adaptive approach to planning, social housing in Vinh was, arguably, one of the first ecological complexes in Vietnam, though it would hardly be remembered as such today. Socialist urban projects such as these are often considered "failed experiments" by international experts working on urban development projects – including experts connected with UN Habitat and German government organizations such as GIZ – and the green standards they have set as the benchmark for contemporary urban development. While the Vietnamese and East German planners I interviewed agreed in hindsight that the complex could be seen as an earlier iteration of

38 Personal interview with author, Vinh City, January 17, 2011.

green architecture, they stressed that this nomenclature was not in use at the time. Rather, it remains a bourgeois conceit of capitalist urban planning, as I argued in the introduction. There was no consensus, however, regarding what motivated ecological planning: some gave materialist explanations, such as poverty and a lack of resources, while others emphasized environmental sensibilities and the attempt to create climate-appropriate dwellings. Instead, my research shows how material constraints *and* logics of sustainability worked in tandem with one another to create a socialist utopian imaginary of the good life for the working masses. Owing to dire postwar conditions, which contributed to a decrease in living standards in Vietnam over the course of the project, "modern" housing amidst canopies of trees was not enough to bring this good life to fruition. Rather than use green spaces for leisure as envisioned, residents turned them into urban gardens and grazing fields to mitigate against food insecurity.[39] More than 30 years later, when taking an elevated view of the housing complex from a neighboring high-rise condominium, one sees the relic of this utopian planning through the lush expanse of trees and green spaces that enfold the now-decaying buildings (Figure 6).

Figure 6: Bird's eye view of blocks C2, C3 and C4 from neighboring high-rise, 2012. Photo by the author.

39 See Schwenkel, "Eco-socialism."

Postsocialist Trees and the De-Greening of Cities

Trees have produced an unexpectedly green socialist city, and yet these very trees are at risk with current plans for urban renewal. Scholars have shown that political and economic changes often involve new relationships between humans and nature.[40] This has especially been the case with post-socialist cities that have been subjected to capitalist restructuring, as Hirt's analysis of Sofia reminds us. Accompanying the removal of "undesirable buildings" from the landscape[41] has been a reduction in the expanse of public green space. In the infamous "shrinking cities" of East Germany,[42] the death of buildings – those socialist "eyesores" slated for demolition – has also meant the death of nature in the city.

The destruction of green spaces by redevelopment schemes constitutes the second phase of Benjamin's fallen nature, but this time nature has fallen victim to the violence of capitalist accumulation and its privatization of parks, lakes and gardens that have created new zones of segregation.[43] Despite all the trendy "eco" talk one hears today across Vietnam, urban space in Vietnamese cities is by most observations *less* green that it used to be. Not unlike the *Wohngebiete* in Vinh, socialist housing projects in Hanoi, tended to follow an open landscape design in accordance with Soviet planning practices (which had also influenced the GDR). Planners built rows of apartment blocks an average of 24 meters apart to provide ample sun, air, and greenery for residents.[44] Over time in complexes such as Nguyễn

40 See Anne Rademacher, *Reigning the River: Urban Ecologies and Political Transformation in Kathmandu* (Durham, NC: Duke University Press, 2011).

41 Melánie van der Hoorn, *Indispensible Eyesores: An Anthropology of Undesired Buildings* (New York: Berghahn, 2009).

42 Felix Ringel, "Post-Industrial Times and the Unexpected: Endurance and Sustainability in Germany's Fastest Shrinking City," *Journal of the Royal Anthropological Institute* 20, no. S1 (2014): 52–70.

43 Erik Harms, *Luxury and Rubble Civility and Dispossession in the New Saigon* (Berkeley: University of California Press, 2016).

44 Đào Tiến Ngọc, "Chỉ tiêu và thành phần không gian xanh trong khu đô thị mới," *Quy Hoạch Xây Dựng* [*Planning and Construction*] 50: 67.

Công Trứ – the first attempt at standardized housing – communal spaces were cemented over and makeshift structures came to replace trees and shrubs as residents expanded their living space and economic activities beyond their small units. The "cementification" of public green spaces in Vietnam must then be seen as an effect of the market economy, rather than an intrinsic outcome of socialist planning it is often the charge in the West.[45]

Green spaces are reemerging as places for the wealthy in Vietnam – quite ironically given that the country remains under the rule of the Communist Party. In contrast to the socialist past, new eco-scapes built on the outskirts of major cities and sold by upscale developers are increasingly spaces of affluence and exclusion, undermining state policy that proposes the democratic greening of cities in keeping with earlier practices. Yet, according to reports, land allotted for green space in new urban projects is on average less than 0.5 square meters per person,[46] up to five times short of regulations and three times less than the average proposed by GDR planners in the *Wohngebiet* in Vinh. This is because despite promises to the contrary, revitalization has reduced the density of green spaces in socialist cities. As the architect Đào Tiến Ngọc has lamented, because land now acquires market value, investors attach little importance to the inclusion of green spaces in their redevelopment plans and officials rarely enforce urban policy.[47] Under a system where the spatial logics of capitalism have become deeply embedded in the landscape, concerns for private property outweigh concerns about ecology. Residents in the housing blocks in Vinh have similar anxieties about renewal schemes in their neighborhood: that such future designs will reduce the amount of communal green space, despite assurances to the contrary.

The inhabitants of the *Wohngebiet*, which was denationalized in 2011, live on some of the most valuable property in the city today. They

45 Schwenkel, "Eco-socialism," 51.

46 TTXVN, "Urban Area Plans Fall into Disarray," *Việt Nam News*. July 29, 2011. <http://vietnamnews.vnagency.com.vn/Pages/PrintView.aspx?ArticleID=2>, accessed July 2011.

47 Đào Tiến Ngọc, "Chỉ tiêu và thành phần không gian xanh trong khu đô thị mới," 68.

also live in the city's most decrepit buildings, revealing how the labor heroes of socialism, in a postreform society, have become the new urban poor. That the working classes even remain in the city center, however, is a testament to the history of socialist urbanization that I have outlined above, and its unique forms of spatial organization: high-density mass housing. Urban renewal policies in the Global South have resulted in the mass displacement of impoverished communities to the margins of cities – a classic form of accumulation by dispossession that geographer David Harvey has called the new imperialism.[48] In the case of Vinh, however, residents – now legal owners of the apartments – have organized to withstand commodification, or at least to have a say in how privatization unfolds *without* their relocation. In the city's action plan for sustainable growth, the obsolescent blocks are slated for demolition to make way for more spacious condominiums, sold as a modern paradise – market socialism's new utopia. Green spaces once put to good use for the benefit of workers and their families are now lucrative lands in need of development, thus threatening the core principles of eco-socialism that advocated urban nature "for the people" and not "for profit." From a developer's perspective, a retired bureaucrat at the Department of Construction (who himself had lived in the blocks) explained to me in an interview, the residential complex was full of "wasted land" (*lãng phí đất*) because of its open landscape design. This, despite the active use and appropriation of such space as playgrounds, gardens, grazing lands, sports fields, and even small business operations (such as barbershops and tea stalls; Figure 7). "Those green spaces would not exist today if they were to rebuild," the bureaucrat forecasted.[49] In other words, where residents saw public space for their leisure, such as playing soccer, developers saw plots of unused land for speculative development.[50]

48 David Harvey, "The 'New' Imperialism: Accumulation by Dispossession," *Socialist Register* 40 (2004): 63–87.

49 Personal interview with author, Vinh City, May 22, 2011.

50 At the time of writing, the soccer field in Area A is now a construction site between the housing blocks that have yet to be demolished.

Figure 7: Multifunctional, communal green space used for parking bikes, grazing, gardening, drying clothes and playing badminton (a team of grandmothers in the background), 2015. Photo by the author.

This will indeed change in the future as urban expansion threatens to eliminate surplus green space. In one advertisement for renewal in the housing complex, a blueprint depicts a single row of trees that encircles a series of upscale high-rise buildings, with no open spaces for recreation and assembly. This was similar to the case of Block C1, which was demolished in 2004 and replaced by two towers that adjoin an asphalt parking lot. The trees and greenery promised to residents are stunted bonsais growing in cement containers. "This is what the future of development will look like," commented Mr. Ngô Văn Yêm, a retired Vietnamese architect, trained at the *Hochschule für Architektur und Bauwesen* in Weimar, who continues to admire what he sees as the *Wohngebiet's* superior ecological planning with its attention to open green spaces populated with trees and and bright and airy apartments.[51]

51 Personal interview with author, Vinh City, January 11, 2015.

Residents also exhibited such skepticism. Despite sharing a dream of urban betterment, seniors, in particular, have not been persuaded by what they see as deceptive images of a greener future. Indeed, the construction of a towering condominium on a large open field in Area C (adjacent to the twin high-rises) reaffirmed such sentiments. Not only did it destroy the sports ground, they lamented, but it obstructed the path of wind, leaving the neighboring blocks airless and dusty. "It's just chaos (*lung tung*)," Mr. Yêm complained, not unlike the early socialist planners in East Germany who pointed to the disorder of "capitalist cities," and argued that unchecked growth and unregulated planning lowered the standard of living for all but the wealthiest citizens.[52]

These not-so-sustainable settlements have evoked feelings of nostalgia for the ordered rationalities of the East Germany's green design, which my research respondents cast as higher quality and technologically superior to new Vietnamese developments (notably: not foreign invested!) that they saw as inferior in design and construction. These technological enchantments with Germany, which reproduced hierarchical difference as the natural order of things, stemmed from a general distrust of local government and its land-use regulation. While living in the blocks and conducting research, my elderly residents repeatedly told me that new housing projects have not generated the green spaces and pathways, nor the ventilated apartments, like their *Wohngebiet*. For them, the aesthetics of capitalism – not socialism as the literature shows – connoted "greyness" and inhabitability: blocking the sun, disrupting air flows, eliminating gardens, and ridding them of collectively used space that is vital to maintaining their social activities. "East Germany planned this city well," a municipal authority at the People's Committee stated. "It was a design that considered *con người* [humanity]."[53] These kinds of sentiments expressed at the higher levels of government revealed another layer of ambivalence about the capitalist direction of the city (and the country more broadly), even as these officials have the most to gain from neoliberal development.

52 See, for example, Deutsche Bauakademie, "Thesen zur 1. Theoretischen Konferenz der Deutschen Bauakademie," *Deutsche Architektur* 9, no. 10 (1960).

53 Personal interview with author, Vinh City, February 22, 2011.

Conclusion: Eco for Whom?

I began this chapter with the suggestion that capitalism, not socialism, has resulted in the death of nature in the city owing to the undoing of eco-socialist planning, and ended with an example of the de-greening of cities as public space is privatized and removed from the domain of collective and versatile use. I argued that access to urban green spaces has become less democratic and more controlled, favoring the affluent over the poor. As Hirt has observed, the grand public spaces of socialism have lost much of their public-ness.[54] This is not to wax nostalgic about the greener pastures of socialist pasts, but to point to the unique *ecological* characteristics of socialist cities, using the example of East German urban planning in Vietnam to show how trees were part of a history of using urban nature to achieve social stability and build a more equitable city, which is now being undone as public goods turn private. In so doing, my aim was to complicate the conventional association of socialism with the gray dreariness of urban space by showing how postwar Vietnam served a site of green experimentation for East German planners to build socialism through the construction of environmentally sustainable housing for an emerging proletariat. By preventing the privatization of urban nature, GDR planners generated a surprising amount of green space in Vinh that is disappearing with commodification as public goods are turned into private assets for speculation. This is perhaps one of the great deceptions of contemporary green discourses under capitalist restructuring in Vietnam: the flashy ads for beautiful, green living environments are all but a utopian fantasy (or phantasmagoria in Benjaminian terms)[55] propelled by the mass media. There were of course ideological motivations behind the more inclusive designs of the *Wohngebiet*. Socialist planners envisioned parks and other spaces of leisure as restorative for workers precisely to increase their efficiency; in other words, to serve the large goal of socialist nation-building. And the

54 Hirt, "(Post)socialist City," 36.

55 Thanks to Eli Rubin for this point.

GDR's investment in technological worlding cannot be disconnected from its attempt to achieve international legitimacy and to triumph in its Cold War competition with the FRG.

The loss of public space and the death of trees have generated new vulnerabilities, including displacement for some (those who cannot afford new rents) and social death for others. It is a troubling irony that one of the most eminent architects of Vietnam, Nguyễn Trực Luyện, who studied ecological planning in the Soviet Union in the 1950s has become an unexpected casualty of de-greening. This despite (and because of) his family's decision to move into an upscale condominium in the center of Hanoi. The spectacular view from his flat on the eleventh floor served as a metaphor for the simulated world of capitalism far below that was there for the viewing, but only from the margins. During one visit I asked Mr. Luyện how he liked high-rise living with sightlines of the developing city beneath him, surely a sign of the country's prosperity that persons like himself had worked his whole life (and fought in two wars) to achieve. He answered pensively that it entailed the difficult loss of everyday sociability. "There is no place for the elderly to gather," he lamented.[56] Without green spaces close by, he passes most of his time alone inside. In the housing complex in Vinh, on the other hand, seniors spend much of their days outdoors socializing with one another, gossiping and drinking tea, playing board games (men) or badminton (co-ed) in communal spaces. And yet it is precisely the loss of the collectivity as the quintessential experience of this new urban modernity that makes them apprehensive about redevelopment.

Were cities built differently under capitalism and socialism to produce unique characteristics? While many scholars have grappled with this question, I decided to pose it to one of the senior East German architects, Dr. Karlheinz Schlesier, who worked on the master design of Vinh, and who had strong feelings about this issue. They may look the same, he told me, referring to the construction of mass housing under both systems, but they served different purposes. And yes, he answered, those in the East were considered greener. His view emphasized the value of public goods,

56 Personal interview with author, Hanoi, January 10, 2015.

including nature, for the social, material and physical betterment of society: "Residential complexes were built with people in mind to provide the best living situation possible, with parks, fresh air, and sports. This was a major difference: capitalist planning placed value on land, while socialist planning emphasized the needs of people, what they required to be productive citizens."[57] As his words remind us, green planning was as much as about creating a specific type of person – productive, healthful socialist ones – as it was about shaping the environment.

Members of the older generation in the *Wohngebiet* in Vinh hold tight to such ideological commitments as expressed by Dr. Schlesier. Despite the massive decay, they attempt to retain the prestige that once came with their eco-friendly apartments. The younger generation is more ambivalent, however, as green fetishization often signifies their exclusion. One need only quickly thumb through the glossy architecture magazines now popular in Vietnam to see how eco-creations are deceptively beautiful, upscale, and decidedly *not* for the masses – a quick transformation in a short period of time. Greenness has thus taken on an entire set of new meaning and values to uphold a particular fantasy in contemporary urban settings. "Vietnam doesn't have the luxury to care about green cities," explained Mr. Nguyễn Văn Tài, a younger Vietnamese architect born after the war, who had an entirely different relationship to ecological planning – one that was informed by capitalism and green space as exclusionary. "People are more worried about jobs, housing, and getting their kids into good schools."[58] As his response shows, the new green is not about "the people" (*nhân dân*) as it was for previous generations, but rather serves as a marker of social and economic difference. As such, his critical stance raises the important question that the public in Vietnam is starting to voice more loudly as urban space is progressively stratified, privatized, commodified, and de-natured: eco for whom?

57 Personal interview with author, Berlin, August 29, 2012.

58 Personal interview with author, Hanoi, December 12, 2011.

KATRINA NOUSEK

"Zweige, Nadeln, Dreck": Dwelling on the Social in *Simple Storys* by Ingo Schulze

At first glance an unlikely story about nature, Ingo Schulze's novel *Simple Storys. Ein Roman aus der ostdeutschen Provinz* (1998) sets readers amidst dwindling employment opportunities in Altenburg, Germany, and economically motivated moves to Leipzig, Stuttgart and a recently reunified Berlin. Its figures negotiate professional and personal relationships imbricated with the incursion of new corporations, shifts in political affiliations, and refugees fleeing the Yugoslav wars in the years after the official end of real existing socialism in Europe. However, as the 29 chapters of the novel unfold, they offer readers intertwining vignettes of political grudges and romantic intrigues that draw heavily on natural imagery while alternating between third- and first-person accounts of the time. Shifting from figure to figure, these short episodes give readers a multiple and sometimes contradictory perspective on events that happen not in Berlin after the *Wende*, a location that has since become representative of postcommunist Germany,[1] but rather in the Thuringian "ostdeutschen Provinz" of former East Germany from which the novel gets its name.

1 The terms postcommunism and postsocialism are sometimes used interchangeably to describe eastern European areas transitioning from socialist states to capitalist forms of government. Whereas postsocialism has a much wider field of reference, beginning during Latin American revolutions in the late 1970s and extending through eastern Europe and the Middle East to present-day China, political theorists and economists in the West initially used the term postcommunism to refer to the drastic imposition of economic paradigms of transition that ignored legacies of real-existing socialism and other differences among the states subjected to them. Eastern European artists and intellectuals have purposefully reworked this

Two constellations of figures are central to the novel: one involves Barbara Holitzschek, a doctor at the psychiatric ward in Leipzig-Dösen, whose partner Frank is a politician from former West Germany in the Green Party. The other constellation of figures involves Martin Meurer: the son of Renate, a secretary who lives in Altenburg-Nord, and the stepson of Ernst, a school principal and former member of the Communist Party embroiled in accusations over a colleague he fired. Martin, an art historian who loses his position in Leipzig when the faculty changes in the early 1990s, ends up in Stuttgart at the conclusion of the novel, where he works for *Nordsee* distributing advertisements while dressed in a frog suit.

Most scholarly readings of Schulze's novel have focused on the author's use of the short story form adapted from Raymond Carver and Ernest Hemingway, as well as filmic techniques and an episodic structure they attribute to Robert Altman's cinematic adaptation of Carver's stories in *Short Cuts* (1993). These readings show how American culture productively shapes the author's work despite its critical view of Western capitalist influence (and Western form more broadly speaking).[2] Though these readings are not strictly biographical, they point to the author's adeptness at adopting

term to demonstrate the aesthetic potential and historical particularity of eastern European cultural productions. Within German intellectual circles, philosopher and media theorist Boris Groys influentially named the cultural politics and economic disenfranchisement specific to unified Germany and the former Eastern Bloc the "postcommunist condition." See the proceedings of Groys' 2004–5 research program published in Boris Groys, Anne von der Heiden and Peter Weibel, eds., *Zurück aus der Zukunft: Osteuropäische Kulturen im Zeitalter des Postkommunismus* (Frankfurt/Main: Suhrkamp, 2005).

2 See for example Christine Cosentino, "Ingo Schulzes 'Simple Storys' und die Tradition der amerikanischen Short Story," *Germanic Notes and Reviews* 31, no. 2 (2000): 134–138. Friedhelm Marx emphasizes the literary influence of both Hemingway and Carver while clearly distinguishing Schulze's writing from postmodernism. According to Marx, Schulze channels the multiplicity of perspectives that sometimes marks postmodern literature through a single proper frequency that creates a "Poetik der Angemessenheit" instead. Friedhelm Marx, "Die 'richtige Frequenz': Ernest Hemingway und Raymond Carver in Ingo Schulzes Roman *Simple Storys*," *Zeitschrift für deutsche Philologie* 124, no. 4 (2005): 595–608. Here 608.

formal strategies that help him capture the so-called postcommunist condition in rural areas of Germany to which he also belongs. Schulze, born in Dresden, East Germany in 1962, made his literary debut in 1995 after the unification of East and West Germany. He published *Simple Storys*, his second novel, three years later after half a year spent in New York City studying short stories. In one of the earliest analyses of Schulze's novel, Christine Cosentino notes Schulze's unique ability to imitate and personalize the literary strategies characteristic of short stories, such as telling silences, scant psychologizing and direct speech.[3] Reiterating Ulrich Greiner's review from *Die Zeit*, Cosentino locates Schulze's success in his use of a foreign literary form that successfully captures the alienation of rural life in postcommunist eastern Germany.[4] In a similar vein, Boris Groys emphasizes the critical distance Schulze achieves through formal imitation, which distinguishes his works from more moralistic and nostalgic accounts of postcommunism. This disciplined and sober narrative perspective earned the author international attention including the 2001 Joseph Breitbach Prize and reviews in publications ranging from *The New Yorker* to the *Financial Times*.[5]

Yet Schulze's appropriation of literary strategies attributed to western authors is not merely a matter of successful formal imitation. It depends on difference as well. In Paul Cooke's extended postcolonial reading of the novel, former East Germans insist on distinguishing their histories from those of West Germans after unification in order to produce their own subjectivity. Though Schulze's figures may appear on one level nostalgic or reactionary due to their preoccupation with GDR history, their stories combine in the novel to open a space of difference, or "Trotzidentiät," in

3 Cosentino, "Tradition," 134.
4 Cosentino, "Tradition," 134.
5 Boris Groys, "Die Disziplin der Selbst-Distanzierung. Ein Meister der 'condition postcommuniste': Boris Groys' Laudatio auf Ingo Schulze anlässlich der Verleihung des Joseph-Breitbach-Preises," *Frankfurter Rundschau*, October 4, 2001, 4. See Peter Graves' evaluation of Schulze's international reception in "How Simple are Ingo Schulze's 'Storys'" in Arthur Williams, et al., eds., *German-Language Literature Today: International and Popular?* (Oxford: Peter Lang, 2000), 197–206.

which "competing models of narration" reveal East Germany as a site being colonized by Western forms.[6] These non-identical narrative strategies produce East Germanness not as a defensive reaction to change, but as a dynamic sense of selfhood grounded in a shifting, subjectively constructed past. The communal identity that the accumulated stories produce does not efface individual difference into the faceless we of official GDR state ideology, but offers guardedly non-optimistic moments of solidarity and transformation through Schulze's detailed vignettes of individuals who negotiate changing social orders together in the wake of reductively binary Cold War geopolitics.

Cooke's study demonstrates the cultural specificity of East Germanness as a new mode of negotiating uncertain futures after German unification. Though it complicates simplistic accounts of homogeneous Cold War blocs, it overlooks the figures who enter the narrative subtly – from the margins of the margin so to speak. These figures orient Schulze's story world not only toward western culture and the impact of capitalist economic and political forms after the end of real existing socialism in Europe, but also toward subjects migrating to Germany from other countries in the former eastern Bloc. Attention to these figures demonstrates the social potential to be found in the historical rupture producing not merely East Germanness, but also a more fundamental, specifically postcommunist reconfiguration of time and nature that emerges through narrative practices.

In what follows, I will show how natural imagery involving animals and environments brings the figures in Schulze's novel together despite forces of social fragmentation resulting from shifting state structures in former East Germany. My reading is guided in part by Jason W. Moore's theoretical contribution to Marxist and environmental thought, which seeks to read capitalism through the relationality of humans, nature and environments. Moore seeks an alternative to dualisms between humans and nature that instead reads the progression of history through a dialectical relation in which humans are a co-producing part of a "web of life."

6 Paul Cooke, "Beyond a *Trotzidentität?* Storytelling and the Postcolonial Voice in Ingo Schulze's *Simple Storys*," *Forum for Modern Language Studies* 39, no. 3 (2003): 291.

According to Moore, "An alternative [to dualistic Green thought] begins neither with 'humans' nor with 'nature' but with the *relations* that co-produce manifold configurations of humanity-in-nature, organisms and environments, life and land, water and air. 'History,' in this sense, is the history of a 'double internality': humanity-in-nature/nature-in-humanity."[7] Keeping Moore's focus on the relationality of ecologies in mind, I will show the interdependency of a story world that configures the former German Democratic Republic (GDR) not only in light of the confluence of real existing socialism and capitalist liberal democracy, but also as a site of migration from postcommunist countries that reconfigures social structures for a postcommunist future.

In the novel's fifth chapter, entitled "Zugvögel," Dr. Barbara Holitzschek arrives at the *Naturkundemuseum* in Altenburg where Lydia, a recent arrival from Berlin, has just started a position as taxidermist. Barbara has ostensibly run over a badger, which the museum can use to replace a previous specimen that has gone rotten in an unplugged refrigerator. Lydia accompanies Barbara to retrieve the badger, and most of the chapter follows the conversation the two figures have in Barbara's car. As readers find out, the accident with the badger is probably an alibi, and Barbara has likely hit Martin's wife on her bicycle. Narration surrounding the event remains oblique, however, and is reported largely through affective cues (Barbara's untimely blush and nervous manner) and other figures' suspicions. The narration becomes increasingly unreliable when Martin and his mother date the event to two separate years (1991 and 1992),[8] and when Barbara's traumatic dreams about it are soothed by her husband Frank, who also insists it was a badger.[9] In Chapter 5, the accident has just taken place, and Barbara's nervous hurry emerges in a conversation between her and Lydia that dwells on environmental concerns ranging from the dangers of animal captivity to the effects of climate change. The chapter, conversation

7 Jason W. Moore, *Capitalism in the Web of Life: Ecology and the Accumulation of Capital* (London: Verso, 2015), 5.

8 Ingo Schulze, *Simple Storys: Ein Roman aus der ostdeutschen Provinz* (Munich: dtv, 2006), 104 and 233 respectively.

9 Schulze, *Simple Storys*, 189.

included, is reported through Lydia's first-person narration and structured by the duration of her workday: she is at her desk on a Monday when the chapter begins, interrupted by Barbara's frantic offer of a desirable specimen, and returns to resume work when it ends.

Lydia and Barbara have very different social and professional relationships to life in its inanimate and animate forms. Whereas Lydia spends her working hours stuffing animals for display and takes a hardline attitude on her mother's difficulty turning away stray cats destined for death, Barbara admits that her squeamishness around wounded lifeforms led her to a career in psychiatry despite her interest in animals. When Barbara recalls a significant moment from childhood, however, her aversion turns out to be rooted not exactly in animals, living or dead, but rather in a traumatic memory involving a shred of artificial fur and a conversation with her mother: "Wenn ich mich gruseln will, brauche ich nur daran zu denken, an das verformte graue Ding mit den Fellfetzen ... Ich sollte ihr [der Mutter] nie wieder damit kommen, daß ich was mit Tieren machen möchte."[10] The grey thing, a first-aid kit that melts to the fur-covered backseat in her family's Skoda, is hardly an animal, yet in the conversation it functions to produce an association between Barbara and Lydia in which the latter connects her narrative to the former through an artificial thing resembling an animal. The imagery, though Barbara associates it with animals, is not itself natural, but rather a synthetic material made to resemble nature. As a narrative function, however, this matters less than the metonymic bridge it forms between the figures, whose desires, dreams, and professional futures turn out to be bound up in affective reactions to animal, or animal-like, things.

Though Barbara's disgust at the artificial fur has kept her from working with animals professionally, she seems to have a vested interest in animals that "man muß ... wie Menschen behandeln."[11] In the course of conversation, she relates a series of facts about the digestion system of komodo dragons, stress that builds in captive dolphins, and monkeys that eat their young. Some of these facts she has gleaned from *Der Spiegel*, and her knowledge is likely also related to her husband Frank, a politician in the Green

10 Schulze, *Simple Storys*, 53.
11 Schulze, *Simple Storys*, 54.

Party. Barbara's nervousness, focalized through Lydia's perspective, takes the form of constant chatter interrupted only occasionally by Lydia's pragmatic interest in finding the badger and getting back to work. Barbara's display of knowledge hardly impresses Lydia, who serves first as a social point of orientation for Barbara's conversation, and later as a physical anchor when she takes over the wheel for the return trip: "Sie [Barbara] verläßt sich völlig auf mich."[12] In so doing, Lydia loses her own sense of orientation when she drives back and must depend on Barbara's directions home.

While Barbara recapitulates facts about the environment, Lydia works practically with animals in order to preserve them as dead exemplary forms. As a taxidermist, Lydia quite literally objectifies animals for work, which inverts Barbara's obsession with the artificial fur in the Skoda that reminds her of animal life despite its synthetic production. Barbara's generalizations about the large-scale movements of species give readers a scientific perspective from a bird's eye view. These are not individual specimens, but generalizations about groups that recognize the interconnectedness of species and environment. However, despite Barbara's broad display of facts, she has little familiarity with how particular animals behave. The reader's first tip that something is afoot behind the badger incident, for example, occurs when Barbara imitates the badger she has hit, only to demonstrate holes in her knowledge – and her alibi – when Lydia skeptically remarks, "Das soll ein Dachs sein?"[13] Barbara lacks the practical knowledge Lydia gains from handling individual organisms at work, resulting in a creaturely imitation that bears little resemblance to an actual badger.

In a key part of the dialogue, Barbara takes the warbler Lydia was stuffing when they left the museum as a point of departure to describe her wonder at shifting migratory patterns in birds. The scattering of facts and statistics she relates are summarized for readers succinctly through a series of ecological patterns that connect the speed and stamina of the birds to a sequence of large-scale, structural changes in the environment: "Danach spricht Dr. Holitzschek von der Erderwärmung, dem steigenden Meeresspiegel, der Versteppung von Rastplätzen, von schwindendem Futter

12 Schulze, *Simple Storys*, 58.

13 Schulze, *Simple Storys*, 51.

und der Verschiebung der Jahreszeiten."[14] After this summary of ecological crisis condensed into a single sentence, the narrative shifts focalization back to Barbara, who then muses on the *Zugvögel* of the chapter's title. Due to global warming, some birds have ceased migrating at all, and those that continue to migrate find their nesting grounds occupied by common, more aggressive species: "'Was wirklich schlimm ist,' sagt sie, 'wenn die Nachtigall oder der Pirol in ihr Brutrevier kommen, sind die besten Plätze längst von Allerweltsvögeln besetzt. Die sind aggressiver und ausgeruhter als die richtigen Zugvögel.'"[15] Though Barbara is troubled by the competition for breeding grounds that global warming produces, her remarks, presented as direct speech in the narration, never mention the effect of human impact involved in producing global warming as such. Instead, while Lydia watches Barbara speed past container trucks and cattle carriers that are presumably contributing their emissions to the problem, Barbara wonders at the "ineinandergreifende Orientierungssysteme"[16] that the birds use to find their path each year. The ecological picture she paints is not one of human intervention in a nature separate from it, but rather a set of interlinked patterns, groups of organisms, and systems that both changes over time and embraces scientific intervention as a manner of balancing antagonistic relations. Notably, Barbara follows her above-quoted lament about the difficulty nightingales and orioles have finding breeding grounds with a pragmatic-minded interest in genetic manipulation that might create new species through "Kreuzungsversuche von Stand- und Zugvögeln."[17] Instead of lamenting the loss of nature climate change produces, Barbara focuses on the dynamic aspects of these organisms that are living parts of the environments making them, be it through inner clocks that shift with the shifting seasons, or the potential of producing more resilient species through cross-breeding.

The need for a new, interdependent system of orientation that adapts to shifting climates serves as a cipher for the figures' movements in the

14 Schulze, *Simple Storys*, 55.
15 Schulze, *Simple Storys*, 56.
16 Schulze, *Simple Storys*, 55.
17 Schulze, *Simple Storys*, 56.

wake of political changes in their own environments. Many of these figures have not moved far, if at all – Altenburg, Leipzig, Berlin and Stuttgart are moderate distances compared to those a figure from the former Yugoslavia travels – but the social and political climate has moved around them. The economic changes foisted upon the characters during the transition to capitalism and western-style liberal democracy contort social relations, bringing back feuds and intrigues from the past. The opening chapter, for example, introduces readers to a long-standing argument between Dieter Schubert, a former school teacher, and Martin's stepfather, Ernst Meurer, whom Dieter addresses pejoratively as the "roten Meurer."[18] As readers later learn through bits and pieces of their story that appear gradually throughout the novel, Ernst fires Dieter according to Party orders in the late 1970s for allegedly tolerating anti-Bolshevik acts, and then Ernst loses his own job with a newspaper after it published an article accusing him of corruption. When Ernst circulates a letter in his defense that fails to garner support, he begins a retreat from public life that will eventually lead to his psychological decline and internment in the Leipzig-Dösen psychiatric ward under Barbara's care. Dieter's professional path takes unexpected turns as well, leading him through a three-year probationary stint in a coal factory and eventually to study museum education before dying from a heart attack while fishing. These relationships are fraught with corruption and betrayal, yet they also underscore the social interconnectedness of the figures as they adopt and adapt political and economic forms in the wake of real existing socialism. Most importantly, the figures' interdependent lives – connected formally through their intertwining narratives and thematically through images of nature marked by human construction and artificial production – suggest the necessity of finding new points of orientation articulated through the "web of life."

In *Capitalism in the Web of Life* (2015), Jason W. Moore argues for a non-dualistic relation between nature and society that understands capitalism as a historically specific part and parcel of a dialectic between humanity-in-nature and nature-in-capitalism. Moore writes,

18 Schulze, *Simple Storys*, 21.

> While the manifold projects of capital, empire, and science are busy making Nature with a capital "N" – external, controllable, reducible – the web of life is busy shuffling about the biological and geological conditions of capitalism's process. The "web of life" is nature as a whole: nature with an emphatically lowercase *n*. This is nature as us, as inside us, as around us ... Put simply, humans make environments and environments make humans – and human organization.[19]

In Moore's combination of Marxist and environmental thought, capitalism is one specific configuration of human relations that is produced not in opposition to Nature, but rather as a world ecology that co-produces nature as part of the historical process. Moore uses this argument to radically rethink Marx' value theory by noting the influence of environment-making processes that are also a part of the capitalist world ecology at a given point in history. The contemporary organization of nature, he argues, is structured by capitalism to find and exploit unpaid work (such as women's work, colonial labor, or labor performed by animals) and to produce "Cheap Nature" in the form of food, labor-power, energy, and raw materials.[20] Far from evidencing a clear dichotomy between Nature and Society, the environments constructed in and through capitalism are shot through with social processes woven from the web of life where class struggle, symbolic meaning-making and limited sources of organic and inorganic material mix. From this view, readers see how capital produced Nature as the symbolic externality to capital's own logic of appropriation, thereby making what stands behind the signifier measureable and open to appropriation. Rethinking capital according to Moore involves seeing humans – as the only species that establishes value as a way to prioritize environment-making human relations – as crucial actors in redefining capital. Given the dependence of capital on finding and exploiting unpaid work, often at the expense of cheaper natural resources and extra-human labor sources, human activities may also establish new relations and make new environments, remaking capital in so doing.

The figures in *Simple Storys* carve out their living and working spaces in the wake of the Cold War, where unpaid work abounds as individuals

19 Moore, *Capitalism*, 3.
20 Moore, *Capitalism*, 53.

rebuild their lives and find what jobs they can despite credentials unrecognized by new employers, and intrigues colored by past involvements with the Socialist Unity Party of Germany. Barbara and Lydia are respectively lucky in this transition: Lydia's training in Berlin has positioned her well to receive full-time work in the more rural Altenburg, and Barbara has a stable position in the psychiatric clinic as well as a husband who works for the new government. Despite their economic successes, however, Barbara entangles Lydia in the bicyclist's death[21] while also acknowledging an environment with migratory patterns that remind readers of the shifting geopolitics and people in recently unified Germany.[22] In response, Lydia takes over the wheel for the return trip to navigate a safer way for the two, developing an integrated system of orientation that relies on Barbara's knowledge in place of her own.

Yet Lydia's actions in this moment are largely decided by her preoccupation with work. She is in the car in search of a specimen for the museum; she is scheduled to lead a school tour later that morning; and, despite Barbara's troubling story, she returns immediately to her desk after the incident, "rück[t] näher heran und arbeite[t] weiter."[23] Because the novel is focalized through Lydia's first-person perspective, Barbara's narrative becomes temporally structured by Lydia's workday, too. Barbara's

21 The alleged death is a crucial structuring element in the narrative, which begins in February 1990 but seldom references temporal coordinates directly. Although Martin dates his wife's death to October 1991, Martin's mother, Renate, recalls the year as 1992 when recounting events to Barbara. Peter Graves chalks this up to an authorial mistake, but given Barbara's status as one of the few figures who has close associations to Western Germans (through her husband), as well as Schulze's care in composing such an intricate novel, I tend to read the mistaken date as a marker of collective confusion during the postcommunist period. Barbara's narrative is out of joint, and the rupture in her experience echoes through chronologies emerging in the others' stories too. See Graves, "How Simple," 201.

22 Ulrike Bremer reads Chapter 5 as an allegory for East Germany undergoing the economic strategy of "shock" privatization after the end of the Soviet Union. See *Versionen der Wende. Eine textanalytische Untersuchung erzählerischer Prosa junger deutscher Autoren zur Wiedervereinigung* (Osnabrück: Universitätsverlag Rasch, 2002), 226.

23 Schulze, *Simple Storys*, 61.

accident will reappear in dialogue with other figures later in the novel, but initially her nervous search for an alibi, and the environmental crises upon which she dwells in the meantime, are fit into the rhythms of capital and the sense of continuous work Lydia establishes. In her rush back to the museum, Lydia not only forgets to give Barbara her key, but also inquires no further to figure out what actually happened. When a colleague asks about the badger, Lydia holds up a very different sort of non-animal, a koala keychain, which has come to stand in place of her intended museum acquisition: "'Den hab ich vergessen', sage ich und halte den Schlüssel [von Barbara] an dem gelben Koala-Anhänger hoch, als könnte ich damit etwas beweisen."[24] What exactly Lydia might prove with the incident is left purposively ambivalent: whereas Lydia's turned back presumably leads her to believe Barbara refers with "den" to the masculine substantive for badger (*der Dachs*), Lydia remains preoccupied with her work while referring to a different masculine substantive: the key in her hand (*der Schlüssel*).

As part of a "web of life" Barbara and Lydia's interactions seem oddly oriented to and through images of death in nature. From taxidermied animals that preserve dead life and the inanimate koala keychain to the mutilated forms of animal-like shredded artificial fur and the dead badger or bicyclist, these figures display a troubled environment inextricable from human intervention, creative and destructive alike. The world in this chapter is constructed under the auspices of Barbara's need for an alibi and through her nervous recapitulation of discourses of environmental crisis and transformation. As a cover story, these discourses seem to fail, resulting neither in a badger for the museum, nor a sufficient solution for Barbara's trauma – not to mention the dead bicyclist. Upon closer reading however, they reveal a world ecology constructing political and physical environments against a backdrop of museal preservation in the wake of an individual's trauma and social rupture more broadly. This is furthermore carefully coded in forms that destabilize easy distinctions among the symbols, materials and activities that organize environments. These human and extra-human natures, to speak with Moore, are used to form narrative connections that rework

24 Schulze, *Simple Storys*, 61.

environments and facilitate social orientation in the absence of political structures that might otherwise provide guidance. Although Schulze describes the large-scale political changes after the end of East Germany only as they affect each figures' experience in the novel, the minutiae that compose the figures' everyday betray a need to reestablish human relations and reassert connections through newly adopted modes of capitalist production. By way of formal qualities that depend on multiple focalization, Schulze establishes a network of narrative connections in which the figures' lives intersect to construct points of orientation through *indeterminacy*.[25] More specifically, Barbara and Lydia's contrived consensus about the badger or key to which they referred is an example of narrative ambivalence in which a single, unified meaning is not determined. Instead, a signifier bears multiple, co-existing meanings that change at different points and in relation to different figures in the narrative.

Later in the novel, another moment of indeterminacy occurs in which a set of figures collects life and seeks to preserve not dead, but living forms while constructing a new dwelling in former West Germany. In the twenty-first chapter, entitled "Zweige, Nadeln, Dreck," former art historian Martin Meurer renovates an apartment in Stuttgart around 1997. He has enlisted the help of Tahir, a Bosnian refugee whose father was killed in the town of Brčko. The timeframe of the novel and Tahir's reference to his father's death suggest that he is part of a wave of refugees coming to Germany during the wars that ensue as the formerly socialist Yugoslavia breaks up into its constituent republics with democratic elections in 1990.[26] Tahir's story

25 Roman Ingarden and Jacques Derrida have influentially established linguistic indeterminacy as an effect of description in literary texts (Ingarden) and the slippage between signifier and signified (Derrida). A feature of the discourse (how a story is told), it is produced through everything from gaps in what is narrated to contradictory accounts from multiple focalizers. For a brief account of the wide variety of uses of this term, see Emma Kafalenos, "Indeterminacy," in *The Routledge Encyclopedia of Narrative Theory*, eds. David Herman, Manfred Jahn and Marie-Laure Ryan (New York: Routledge, 2005), 475–478.

26 In 1992, Germany admitted 1,219,348 asylum seekers, ethnic Germans from the former eastern Bloc and refugees. Among them, 350,000 from Bosnia were granted "temporary protection." As of 2007, 80,000 Bosnian refugees remained in Germany after the

remains vague, but readers learn that his mother is also in Germany, as well as another woman, Fadila, whom Tahir has a vested interest in marrying off to Martin. Though Martin gently rejects Tahir's offer because Tahir formerly referred to Fadila as his wife, he is sympathetic to Tahir's troubles and offers him food while relating his own story of economic disenfranchisement. Together, Martin and Tahir rebuild their urban environment in a scene that establishes narrative connections among the East Germans who dominate Schulze's novel with figures from the former eastern Bloc more broadly.

At the opening of the chapter, Tahir triumphantly enters Martin's apartment with three fish that he has saved from a smashed aquarium and collected in a Bonaqua water bottle. Tahir explains, "Jemand hat Aquarium zerschlagen, großer Streit zwischen allen."[27] The motivation behind the argument and the destroyed aquarium remains unclear, but left with the fish, Martin supports Tahir's efforts to save them and transfers them to a bucket where they can be fed. At the same time, Tahir also becomes a sounding board for stories about Martin's classmates, former academics who were forced out of jobs or into unskilled labor after a new group of professors and students assumed posts at Leipzig University.

The chapter draws suggestive parallels between Tahir and Martin's experience that contrast the economic migration of East Germans with refugees fleeing the wars in former Yugoslavia. Martin's professional demotion from art historian to *Nordsee* advertisement strikes a blow to his dignity (and to his body when he is randomly punched by a passer-by), yet Tahir's troubles put his own momentarily into perspective through references to bloodier revolutions than those in Germany. As the two discuss their pasts, Martin puts Tahir's unpaid labor to work renovating the new apartment. Yet this situation cannot be reductively read as the mechanism of capital

majority were sent back in the wake of the Dayton Peace Accords in December 1995. In 1997 Germany implemented visas for children from the countries that formerly composed Yugoslavia. Deniz Göturk, David Gramling and Anton Kaes, eds., *Germany in Transit: Nation and Migration 1955–2005* (Berkeley: University of California Press, 2007), 504–6.

27 Schulze, *Simple Storys*, 221.

creating and exploiting new sources of labor. Working together also allows the two figures to mark their own temporality as they organize a new environment by reworking the living space. Heeding neither the workday of an electrician who does not show up, nor the ceaseless piling of needles to be swept from the roof, the two work through the web of life to reconstitute human relations in a capitalist world ecology that demonstrates human agency in the construction of a new, specifically postcommunist moment in world history.

Tahir's difference from figures coming from the former GDR is marked with grammatically incorrect German that allows for a multiplicity of referents to play into the narrative construction of natural and urban space in the scene. Later in the figures' exchange, the two men clear pine needles that have accumulated on the apartment roof. While they are doing so, it begins raining, yet readers are only gradually made aware of the turn in weather. The narrative configuration of the scene depends on a play of meanings in *regnen*, which signifies both raindrops on the apartment's roof and the needles that Martin clears from it. While they are working, Tahir remarks twice, "Es regnet," an allusion to the rain outside that first seems to refer to dirt and needles coming from the roof.[28] This initial referent is established through focalization that shifts to Martin, who reports: "Ich laß es Dreck und Nadeln regnen. Wirst sehn, wie hell das noch wird. Das ist das I-Tüpfelchen aufs Ganze."[29] Because earlier in the scene Tahir's grammatically incorrect German prompts Martin to correct him or ask for additional clarification, the *regnen* in this part of the scene is difficult to place until additional information – Martin's wet hair and his mention of an open window that might let rain in – locate it more exactly.[30] Eventually, the third person narration differentiates the weather from the raining needles at the end of the chapter when the actual rain overcomes the other noises: "Martin blinzelt aus der Hocke hinauf zum Dach, auf das, alle anderen Geräusche übertönend, der Regen klopft. Eine Kiefernnadel zuckt zur Seite, ein Stück nur, und wieder zurück, hin und her springt die

28 Schulze, *Simple Storys*, 225 & 226.

29 Schulze, *Simple Storys*, 226.

30 Schulze, *Simple Storys*, 226.

Kiefernnadel unterm Regen, und gleich daneben eine andere Nadel und noch eine und noch eine."[31] Whereas narrative ambivalence is maintained in the scene with Barbara and Lydia through multiple antecedents accruing to the pronoun *den*, here the referent of *regnen* is finally determined. The chapter seems to end on a note of resignation as the raindrops bring more pine needles to cover the freshly cleared roof, and Martin and Tahir's efforts are made futile in the face of natural cyclicity.

Lest readers think the scene simply signals either man dominating nature, or natural cycles that subsume man, however, literary comparisons to a clock and a fish that conclude the chapter return readers to the world ecology the narrative creates among the figures by way of reference to their environments. These literary configurations depend on narrative ambivalence in which natural imagery establishes the temporal coordinates by which the figures orient themselves together. As we will see, the temporality constructed is not a matter of absolute temporal coordinates, but rather a rhythm that plays through Martin, Tahir, and the environment they construct beside their neighbors. Reestablishing a sense of time in narration focalized through Tahir and Martin produces a social bond in the web of life that shifts readers' attention away from the incursion of capitalism and the traumatic experiences of postcommunist transition. Instead of dwelling in feuds from the past, the chapter insists on constructing futural modes of postcommunist social organization oriented toward shared spaces and appropriate to a story world in flux.

The needles falling on the recently cleaned roof suggest a new temporality, or a new configuration of naturalized cyclicity, after 1989. This year marks not only the *Wende* in Berlin, which has caused critical classifications of Schulze's novel as "den langersehten Roman über das vereinigte Deutschland," but also other revolutions and shifting geopolitics throughout the aligned and nonaligned states of the eastern Bloc.[32] With figures like Tahir, Schulze's novel draws attention to the position of the former GDR

31 Schulze, *Simple Storys*, 227.

32 Wolfgang Höbel, "Ingo Schulze: 'Simple Storys' – Glücksritter auf Tauchstation," *Der Spiegel*, March 2, 1998, accessed March 30, 2017, <http://www.spiegel.de/kultur/literatur/rezensionen-ingo-schulze-simple-storys-gluecksritterauf-tauchstation-a-27269.html>.

as a ground where both processes of westernization, and also the broader context of migrations from postcommunist states to the east become visible. When read through the lens of world ecology, this context becomes crucial for seeing the creative aspects of actively shifting patterns in the web of life that Barbara and Lydia's relations seemed limited to preserving in fixed terms of death and crisis.

Tahir opens the chapter with a playful "Veni, vidi, vici," while waving his rescued fish that ironically figures him as a conqueror of nature despite his worn clothing and modest income from playing chess.[33] As soon as he appears on the scene, the narrative foregrounds temporal motifs. A ticking sound which has prompted Martin to call an electrician resonates throughout the apartment, and Tahir's wristwatch is frozen on a second: "Der Sekundenzeiger zuckt hin und her."[34] Time, it seems, has both stopped and become uniform and ubiquitous. Through these narrative conceits, the figures construct their own temporality as they reorganize their environment through narration while renovating.

Because the chapter is dominated by dialogue, readers perceive the space around Martin and Tahir largely through subtle remarks that help them imagine the scene. This narrative strategy makes the ubiquitous ticking sound difficult to place. First Tahir confuses the "tick-tick-tick"[35] for a ticking clock, yet by the conclusion of the chapter the noise dominating the scene shifts to the "tick-tack, tick-tack"[36] of needles continuing to pile onto the roof that Martin has just cleared. Oddly enough, the tick-tack at the end of the chapter – the typical onomatopoeia associated with clocks in German – refers not to the clock, but to the motion of needles gathering on the roof, which is then compared to the fish from Tahir's bottle: "'Siehst du das nicht? Tick-tack, tick-tack.' Martin bewegt seinen Zeigefinger. 'Ja', sagt Tahir, 'zuckt wie Fischlein.' Er lehnt sich an den Türrahmen. 'Wann kommen Elektriker? Wartest du?'"[37] This noise, coded in terms of tempo-

33 Schulze, *Simple Storys*, 216.
34 Schulze, *Simple Storys*, 222.
35 Schulze, *Simple Storys*, 217.
36 Schulze, *Simple Storys*, 227.
37 Schulze, *Simple Storys*, 227.

ral onomatopoeia, establishes a rhythm that changes through the figures' interactions with nature outside the apartment that structures the chapter in relation to natural imagery and the anticipated appearance of the electrician. When the fish Tahir rescued in the beginning of the chapter return at the chapter's end, they become a simile for the moving needles that the figures watch twitching "wie Fischlein" as they settle.[38] The preserved life comes full circle: the figure of a fish, first an animal living in its bottle and then employed as a simile in Martin and Tahir's dialogue, acquires a formal function at the chapter's end. Configuring new narrative relations of time and environment thus takes place by way of a social configuration of nature that occurs both diegetically and at the formal level of discourse itself.

Yet this ecology extends beyond Martin and Tahir. The microcosm they create as they construct a new living space is also embedded in the larger social unit of neighboring apartments being renovated in the former Kindergarten. In the courtyard garden below, another scene of humanity-in-nature appears that offers a less symbiotic relation between urban and natural elements. Thomas Steuber, the current owner of the building, has rescued it from demolition by the previous landlord. Among its unique features are the surrounding fir trees (responsible for the falling needles) and a patch of moss in the courtyard that are rarities in the area. Martin remarks, "Kiefern sind in der Gegend ganz selten. Das Moos auch."[39] The moss Steuber cultivates is therefore "sein größter Stolz," yet he cares little to maintain the gutters filled with pine needles above.[40] Taking matters into their own hands, Martin and Tahir unfortunately clean up the needles by dusting them onto Steuber's children's toys that are lying in the garden. A woman's voice complains about the filthy toys, throwing them onto the moss, which results in Steuber's fit of rage as he hurls the children's things back to his apartment and over a fence: "'Doch nicht aufs Moos!' brüllt er und hebt die ausgestreckten Arme ... Er richtet sich auf, holt aus und schleudert den Traktor auf die Verandatreppe, dann den Kipper. Er schmeißt alle anderen Sachen hinterher, greift sich die Sandform und wirft

38 Schulze, *Simple Storys*, 227.
39 Schulze, *Simple Storys*, 219.
40 Schulze, *Simple Storys*, 225.

sie über den Zaun."[41] For Tahir, Steuber's actions seem disproportionally enraged with respect to the situation, yet they reveal the intensity of his protectionist attachment to what nature remains amongst the buildings, as well as the intimate connections among the figures and their sometimes competing actions as they seek to inhabit the Kindergarten together.

Readers may be tempted to draw connections between the scene and Barbara's comment about common birds taking the best nesting places before the migratory species arrive, yet this model of occupation differs in a few important respects. Steuber, previously the Meurer's neighbor, has connections to western Germany that have likely helped him afford the luxurious ground floors of the building. Yet he has also been instrumental in facilitating Martin's new life during the 1990s. Early in the novel Steuber throws Martin a deal to pick up a BMW in Munich[42] and in this chapter he both gives him an apartment and pays him for help with the renovations. However, Steuber's aggressive actions toward his family's possessions starkly contrast his interest in protecting the moss. Attention to the natural imagery in the scene reveals a new set of attachments undergirding relations among the figures: needles falling in the wrong places bind the figures through a common environment for better and for worse. Just as narrative ambivalence allows a multiplicity of meanings to co-exist in a signifier, these figures find ways of sharing space in light of competing interests.

The showering needles set off a social chain reaction, as it were, and Martin and Tahir's actions directly and negatively affect their neighbors. These are not figures carving out an environment in isolation, but rather interconnected parts of a web of life that encompasses social realms with occasionally antagonistic interests sharing environments. Suggesting the totalizing quality of an environment disinterested in humanity, the rain continues, subduing all noises. Martin and Tahir leave the apartment to enter it and decide not to wait any longer for the electrician Martin had initially hoped would remove the ticking noise. Whereas Lydia and Barbara's interaction had taken place according to the rhythms of capital – an interaction squeezed into the workday – Martin and Tahir take no heed of the

41 Schulze, *Simple Storys*, 227.
42 Schulze, *Simple Storys*, 105.

electrician's hours, refusing to let his absence constrain their movements at the conclusion of a chapter otherwise temporally structured through the wait for him.

The alternation of natural imagery in an urban setting and human efforts at regrowth underscores the interconnectedness of the figures through their environments. Paralleled with the fish in Tahir's bottle, these figures struggle for and against one another to sustain life like colloquial fish out of water among newly reorganized economic and political relations. The social connections articulated through natural imagery bring the figures together in the novel's *web of life*. If on one level the novel exposes feuds and intrigues that accumulated among the figures during the existence of the GDR, on another level, more productive for today, it also carefully meditates on new forms of postcommunist social organization and the need for integrated orientation systems established through capitalist world ecology. When human and extra-human figures in Schulze's narrative are read as co-creators in the web of life, the narrative ambivalence that plays through multiply focalized narration becomes a mode of social orientation. The figures in the novel find their way to one another through narrative techniques that allow them to establish social – and natural – reality in the wake of real existing socialism.

BETTINA STOETZER

Wildes Brandenburg: Engaging "Unruly Nature" in Berlin's Peripheries

"Discover the wild side of Brandenburg", an issue in the local March Oder newspaper invites its readers. Peeking through the dry sandy soil of the region is a lush paradise of species diversity: wolves, great bustards and wild horses lure the reader to join an adventure tour through the Brandenburg "wilderness."

***

Located just 30 kilometers east of the city of Berlin, the March Oder Park is a post-unification era nature park in the state of Brandenburg. Part of the March Oder county district, it is situated in a region that has battled high unemployment and depopulation rates in the aftermath of the fall of the Wall. Founded in the last days of the GDR government, the park has been reconfigured as an eco-tourist destination in the past couple decades. Local media and tourist ads celebrate the region's "idyllic" landscapes and lure travelers to explore a seemingly "uncharted territory." Recently constructed bicycle routes meander over meadows and through deciduous forests, attracting weekend visitors to follow the material remnants of local history and encountering the traces of famous German writers such as Theodor Fontane who depicted the rapidly transforming countryside around Berlin in the nineteenth century. Local forms of adventure tourism such as parachuting, air-ballooning, riding GDR military vehicles, or safari tours abound.

Since the late 1990s, the March Oder Park has also become the residence for many refugees from Africa, Asia and the Middle East.[1] Living in

1 In 2007, an estimated 4,000 to 6,000 asylum seekers lived in 30 refugee homes in different locations in Brandenburg. Since then, numbers have risen significantly

isolated part government-, part private-run shelters on converted military barracks in the forest, many refugee activists, as well as advocacy groups have rallied against these so-called "jungle homes" for many years. Whereas these groups have voiced criticisms of rural racialized exclusions for a long time, it was only until more recently that Germany has moved to the center of public controversy about the "refugee crisis." As the devastating war in Syria intensified and displaced millions of people, many fleeing political conflict in the Middle East sought out Germany as their final destination. In this context, debates about infrastructures of distribution, management and accommodation of asylum seekers have invoked metaphors of chaos and bewilderment encroaching upon Germany: tides, waves, and even tsunamis of migrants flooding uncontrollably into Europe and Germany are seen to put a strain on the country's resources and especially its rural regions, not unlike natural disasters.[2]

This essay draws on ongoing ethnographic fieldwork in Berlin and the March Oder region since 2007.[3] Engaging with interviews and participant observation conducted in the March Oder region, I explore how practices of narrating, performing and embodying the "wild" intersect with a changing politics of race, nation, and class at a time of profound transition.[4] Via several ethnographic vignettes, I attend to the ways in which people

up to almost 36,000 and several new accommodations have opened since 2015. See Gudrun Mallwitz. "Brandenburg hat 73% mehr Flüchtlinge als vor einem Jahr," *Berliner Morgenpost* June 29, 2013, and <https://masgf.brandenburg.de/media_fast/4055/KOMPENDIUM-07-2018.pdf>. In 2015, 1072 asylum seekers were registered in the March Oder Region. See: <http://www.maz-online.de/Brandenburg/Fluechtlinge-Landkreise-sprechen-von-Notstand>.

2 Curiously, these metaphors seem to intensify in media coverage on Germany's rural areas – where refugees are depicted to flood small idyllic villages, see, for example, *New York Times*, November 1 (2015).

3 See Bettina Stoetzer, "Ruderal Ecologies: Rethinking Nature, Migration and the Urban Landscape in Berlin," *Cultural Anthropology* 33, no. 2 (2018): 295–323; and Bettina Stoetzer, *Ruderal City: Ecologies of Migration and Urban Life in Berlin* (Durham, NC: Duke University Press, forthcoming).

4 See also Tracey Heatherington, *Wild Sardinia. Indigineity and the Global Dreamtimes of Environmentalism* (Seattle: University of Washington Press, 2010).

relate to the local environment and invoke unruly nature in order to grapple with conditions of precarity in a zone of "awkward engagement"[5] and encounter. I show how, in their evocations of Brandenburg as a space of wilderness, diverse local actors – ranging from refugees to former East Germans – inhabit an odd border zone[6] in which postcolonial and post-socialist worlds collide.

Recounting several ethnographic stories, I track practices of performing Brandenburg as a wild country: the making of wild bodies and landscapes. These practices echo recent conservation practices and yet they operate across different registers and modes of domination as they emerge in the restructuring of local economies and racialized displacements in a new Europe that is undergoing profound political transformation.

Entering Brandenburg

Leaving the city of Berlin, traveling east, you enter a landscape of pine forests, interspersed with meadows, and agricultural fields. Situated between Berlin and the eastern border to Poland, the March Oder region in Brandenburg is among the most sparsely populated areas in Germany – a borderland of marshes, forests and shifting sands. In an effort to build an economically thriving modern Germany, the region's wetlands were drained by Prussian armies and engineers in the eighteenth and nineteenth centuries.[7] Throughout the twentieth century, two world wars brought violence and death to the region. Toward the end of World War II, thousands of Polish, Russian and German soldiers lost their lives here in what

5 Anna Tsing, *Friction: An Ethnography of Global Connection* (Princeton, NJ: Princeton University Press, 2005).

6 Daphne Berdahl, *Where the World Ended. Reunification and Identity in the German Borderland* (Berkeley: University of California Press, 1991).

7 David Blackbourn, *The Conquest of Nature. Water, Landscape, and the Making of Modern Germany* (New York: W. W. Norton, 2006).

became one of the last battles that would put the fascist regime to a halt. As a result of the war, the region's borders shifted west and the nearby Oder River, together with the Neisse, soon marked the new border between Germany and Poland. As a result, people on both sides of the river were expelled, and the March Oder region became a part of Eastern Germany at the "frontline of the Cold War".[8] In subsequent years, socialist land reform and militarization deeply transformed the region's landscape guarding the borders to the "capitalist West."

After the fall of the Berlin Wall, parts of the March Oder region were transformed into a nature park. In alliance with West German environmentalists, the collapsing East German socialist government created a national park program that would initiate several conservation areas in Brandenburg. Many of these conservation areas had been military zones until reunification. Considered a gift for national unity, the March Oder park was thus among several parks that aided in naturalizing a reunified nation.[9] At the same time, the region witnessed a radical shift in forestry policy and environmental discourse as it aimed towards nature conservation and a mixed forest paradigm ("*Mischwald*") instead of a monocultural approach.[10]

Today, with Poland's integration into the EU, the March Oder region no longer constitutes Europe's periphery but is situated in the center. Yet, since unification, high unemployment rates have plagued the region – at times climbing above 25 percent. As local forestry and agriculture have been restructured, and local industries and military bases were closed down, many residents in the region lost their jobs. In addition, a large percentage of the younger population moved away in the hope of finding work elsewhere. Struggling with depopulation and unemployment for years, the March Oder like other regions in Brandenburg, was thus increasingly portrayed

8 Arvid Nelson, *Cold War Ecology: Forests, Farms and People in the East German Landscape, 1945–1989* (New Haven, CT: Yale University Press, 2005).

9 Today, more than 30 percent of Brandenburg is protected nature and landscape area ("*Landschafts- und Naturschutzgebiet*").

10 For a history of mixed forest ideologies in German forestry and environmental discourse see Simon Schama, *Landscape and Memory* (New York: Alfred Knopf, 1995).

as a place in the backwaters of Berlin – an other Germany, set apart from the West. Images of this "other" Germany proliferated in popular culture, and appeared in film, music and everyday humor and stereotypes. For example, in his song titled "Brandenburg," Rainald Grebe referenced this image: while Berlin had risen as a global metropole where life was flourishing, Brandenburg turned out to be an abandoned place – a transit zone through which people pass, on their way to more exciting destinations. "I feel Brandenburg" (*ich fuehl' mich Brandenburg*), he sings, featuring Brandenburg as a place where things come to an end, where one is bored out of their minds – a place where both a sense of bewilderment and unruly nature have taken over.

This image of wild Brandenburg as a frontier of dwindling populations, unemployment, and unruly natural and social relations, has proliferated across different publics, even though the region has seen considerable economic recovery in recent years and unemployment rates have gone down.[11] Utilizing the "wild" and integrating frontier imagery into strategies of economic recovery and identity formation, many local actors and municipalities advertise the March Oder region as a natural paradise in ads and tourist brochures. Participating in parachuting adventures, horse-riding trips and "Jeep safaris," visitors explore the beauty of the Brandenburg "wilderness. Indeed, with the conversion of almost 30 percent of Brandenburg's territory into nature parks and biosphere reserves, many "wild animals" formerly extinct to the region have found protected habitats in the area.[12]

Faced with growing unemployment and dwindling local economies, many people I met during my fieldwork in Brandenburg developed

11 Reaching their height in the 1990s and then again in 2005 above 20 percent, unemployment rates decreased in the March Oder region since 2008. While they were 14 percent at the end of 2007, they went down to 11 percent in 2010, and have further been decreasing. Yet there are strong differences within the county: ranging from 9 percent in areas with denser population close to Berlin, to significantly higher rates in less populated areas further East. Population density has remained at low levels in much of the county, see <https://lasv.brandenburg.de/media_fast/4055/WEB_BbgSozialindikatoren_2017.pdf>.

12 Jörg Götting, *Wildes Brandenburg: Brandenburg's Nationale Naturlandschaften* (Berlin: L&H Verlag, 2014).

entrepreneurial strategies that drew on these narrative strategies of invoking "wild Brandenburg."

For example, Lars Reichelt, a former car mechanic who grew up in Dresden had lost his job after unification. Inspired by recent efforts to strengthen a sense of local knowledge about the history and nature of the region, he began to organize adventure tours: equipped with a fleet of three Land Rovers, Khaki pants, and a set of binoculars, he offered so-called safari tours, guiding tourists from Berlin and elsewhere to explore the Brandenburg "wilderness." His tours were hodgepodge excursions into present and past landscapes of swamps and forests, haunted by war, and the ruins of GDR bunkers and socialist collectivized farms. Similarly, and yet engaging with a different set of cultural imaginaries, was Carsten's horse ranch a popular stop for weekend travelers, school outings, and "Indian hobbyists."

The "Computer Indian"

Nestled between a large meadow and a small conifer and beech forest, Carsten's ranch is situated at the edge of the March Oder Park as a popular stop for weekend travelers, school outings, safari tourists and "Indian" hobbyists. Located on several acres of land, the ranch features an old ranch building, several stables for horses and livestock, a pasture for 20 horses, a Western town, including a saloon and, sometimes, a small teepee village. Many of the glass windows of the small wooden houses, are broken – from arrow shots. In front of the saloon, you can hitch your horse to a wooden post. Next to the Western town, there is a small square with wooden benches, a big fire pit and a huge stack of hay bales right next to it.

In his early forties, Carsten is tall, has long blond hair, wears glasses and has a friendly smile. When he dons his "Indian" wear, he chucks away his glasses and dyes his hair black. He was born and grew up on the ranch in the 1970s. After World War II, his parents had fled Eastern Prussia – what later became Poland – and settled down in the March Oder region, near the new border. Together with his partner, two children, 20 horses and a bunch of chickens, he now lives on the ranch and organizes "powwows," Western shows, horseback riding workshops and bow and arrow shooting contests. Schools regularly book weekends at the Western town for children to learn about "Indian" wisdoms. "Indian" hobbyists and anyone who is curious come and stay in teepees for a powwow, or they join in workshops to sew traditional clothes or practice bow and arrow shooting skills and horseback riding. Once in a while a safari tour stops by to visit the "wild Indians." A few times a year, Carsten and other hobbyists organize a Wild West show and participate in horseback riding and bow and arrow shooting contests all over Germany, Poland, Sweden and as far as South Korea. Carsten's teepee village is not the only one in the area: in a small teepee village in the park, a local environmental education center regularly offers wilderness camps for city children. Sometimes, his ranch and the center collaborate and exchange ideas. Yet unlike the people at the center, Carsten sees himself not as a professional, but rather someone who "does things for fun," as he explained to me in one of our conversations.

When I first visited Carsten's ranch, he told me that "Indians" had captured his imagination since childhood. During socialism, his parents were religious and did not want him to join any state-run clubs or socialist organizations like the *Junge Pioniere* [Young Pioneers], whose activities included a mixture of outdoor adventure and socialist teachings.[13] Thus, as a child, instead of joining a youth club, he would go to the forest with his friends and play "Indians." But it was his violin teacher who introduced him to the world of "Indians" more thoroughly. Curious, Carsten began to read one novel about "Indians" after another. Finally, after quite some time, he was ready for the famous Karl May novels,[14] which were frowned upon

13 Run by the *Freie Deutsche Jugend* [Free German Youth] the Young Pioneers, served as East Germany's youth organization for children who usually attended "pioneer camps" in the summers and participated in hiking and other outdoor activities.

14 Karl May (1842–1912) was a best-selling author in turn-of-the-twentieth-century Germany. Having written several large volumes on Wild West adventures, his stories revolved around the friendship between the two main characters of Winnetou, an Apache chief, and Old Shatterhand, a white German settler and adventurer. Among

and difficult to find for a long time during communism, because they were considered too pro-Western, too Christian, and most of all, a favorite reading by Adolf Hitler.[15] As some people smuggled them into East Germany, they still carry an aura of secrecy until today. But even more than the Karl May stories, communist "Indians" were Carsten's true heroes: while Karl May movies, focusing on friendship between Native Americans and white settlers, were all the rage in West Germany in the 1960s and 1970s, East Germans created a communist counterpart: the "Red Western," produced by the DEFA film studios, with films such as *Die Söhne der Großen Bärin* (1966), *Spur des Falken* (1968), or *Blauvogel* (1979). These movies were "Westerns with a twist." In them, Native Americans were heroes, and white settlers were capitalist villains. A critique of capitalism and colonialism, the movies depicted Native Americans fighting oppression and expulsion from their land. The main character of the films, played by Serbian actor, stuntman and writer Gojko Mitić became one of the most popular and famous actors in East Germany and across East Europe.[16] After re-unification, he

---

others, the stories include: *Der Schatz im Silbersee* (1890), *Der Schwarze Mustang* (1897), *Der Sohn des Bärenjägers* (1890), *Winnetou I–III*, and *Old Surehand I–III* (1893–6). Although most of the stories are written in first person narrative, from the perspective of Old Shatterhand (as whom Karl May would occasionally dress up), May never visited the American West Coast and only traveled to the Eastern United States late in his life, when he had already written a large part of his work. May also wrote numerous stories about the same character traveling to the "Orient" – although in these stories, he was named Kara Ben Nemsi. See, for example, *Durchs Wilde Kurdistan* (1892), or *Von Baghdad nach Stambul* (1892).

15 Glenn Penny, *Kindred By Choice: Germans and American Indians since 1800* (Chapel Hill: The University of North Carolina Press, 2013), 173.

16 While the so-called East German *Indianerfilme* also highlighted the point of view of "Indians," they stressed conflict over friendship. They thus not only stood in contrast to the West German Karl May movies, but also to Hollywood Westerns that tended to portray the perspective of white cowboys and framed "Indians" as "savages." Movies such as the famous *Sons of Great Bear* (*Söhne der Großen Bärin*), set in East Germany and Montenegro, told a story of corrupt white settlers and fearless Native Americans fighting expulsion from their land. In contrast to Western productions, these films highlighted efforts for "authenticity" (e.g. actors rode horses without saddles). In addition, *Söhne der Großen Bärin* depicted the greed and corruption

performed as the famous "Indian" character Winnetou in the so-called Karl May festival – a re-enactment of Karl May's stories about the American West – near Hamburg for years. Carsten still talked about him with a sense of secrecy and wonder, while lowering his voice, as if he were about to stumble across a treasure chest hidden in the forest.

Yet only after the *Wende*[17] did Carsten begin to be more serious about "Indians." In fact, as he puts it, "re-unification opened up new possibilities for being Indian."[18] In 1992, he and his family had the idea to build the ranch and create a Western space. The land was like a jungle, he told me, and had been overgrown with weeds. They ripped out the weeds, reclaimed the land and made it arable. Gradually, they built the horse stables, then the small "Western Town" and the "Indian" village adjacent to the farm. Soon they began to rear horses, adopting them from people who could not keep the animals anymore.

Carsten remembers the transition in the early 1990s as a radical shift. Before the fall of the Wall, people needed each other. There was a shortage of many goods and possibilities and everyone helped each other out – whether it was fixing a broken car or assisting in house improvement and construction. Soon after re-unification, people changed, according to him. All of a sudden, it seemed to him, people had become stressed and frustrated. Everyone, it seemed, began to hide in their homes behind their TVs. Then Carsten did his civil service year in a retirement home in the new nature

---

of white settlers, broken treaties and the violence of imperialism – but also showed how Native Americans "compromised" to settle and began to use the plough. These representations thus had Cold War undertones and also echoed socialist efforts to create a new worker's state and introduce collective farming. See Thomas Adam, ed. *Germany and the Americas. Culture, Politics and History* (Santa Barbara, CA: ABC Clio, 2005), 550.

17 Commonly used to refer to the collapse of socialism, the beginning of the unification process between East and West Germany, and the transformation from a planned economy to capitalism, the term *Wende* translates as "the turn," or the change.

18 While, in Carsten's opinion, the two Indian hobbyist communities in East and West Germany remained largely separate from each other, there were also a few openings: many East German "Indians" considered western "Indians" to be too commercial but were open to hobbyists who proved otherwise.

park. He recalls his work there as one of the most decisive experiences in his life. Most of the elderly residents of the home found themselves disoriented and confused by the experience of the *Wende*. Their sense of security was shattered. Suddenly, everything that had been their center of life had shifted. Sensing their need for processing and healing, Carsten stopped his routine, and simply sat with the elderly people, listening to them and the stories they had to tell about a life and a world passed by.

"That's why I try to do something else," he explained to me one afternoon, sitting around the fire pit near the teepee village, while his 3-year-old son, Tim, shot arrows against balloons set up in a stack of hay. Carsten insisted he did not miss the former government, or the GDR. And yet, he too mourned elements of the past with, wishing that there had been a different path of re-unification. He missed a time when people "helped each other," and were not so stressed. But as he now knew well, that is not what happened. Carving out his own path in a capitalist society, he eventually became a computer programmer. But while he struggled to find a certain level of self-determination in socialism, this turned out to be an equally elusive undertaking in capitalism. Becoming "Indian" in the Brandenburg "wilderness" thus offered an alternative route.

A self-employed computer programmer today, Carsten wants to be more than simply a worker or a computer nerd. Being "Indian," is thus not simply a name, but it designates a status of being in-between, on the edge of things, and trying things out to see whether they work. "I shoot into all directions in the sky – and I don't want to be pinned down," he said that afternoon during our conversation. Although, he barely got by, he was "a little happy" with what he was doing. He never cared much about money or having a fancy car, and he can't stand being in Berlin – which is too noisy and anonymous. His work is seasonal: during the winter, he repairs computers and fixes PC viruses; in the summer, he travels around and organizes "powwows" and horseback riding events. For him, being "Indian" is thus about being different. His clients and the people with whom he works know him not just as the computer guy, but as "Indian." Hence, they call him the "Computer Indian."

But above all, being "Indian" is about communication. "You don't simply become "Indian" by putting feathers on your head and wearing

Indian clothes," he explained to me. "It's not about how authentic your clothes are, or what "tribe" you belong to. It's about a process of communication. And it's about being in touch with nature, with animals and other humans. For example, it does not matter if the beads on your jacket are in the right order. Instead, you can tell if someone has spent long winter nights stitching the pearl embroidery on one's jacket. You can then tell someone has really engaged."

Carsten sews his own "Indian clothing." In the past, other "Indian" enthusiasts have asked him to sell it. Yet he did not want to alienate his work and turn the artifacts he created into commodities. Instead of obscuring the relationships and labor that went into the production of these artifacts, he wanted to make them come alive. Articulating his own critique of capitalism's commodity fetishism[19] and alienation of labor, he thus invited people to his place and taught them how to sew clothes. "It's about the process of sewing. The same is true for bow and arrow shooting while riding on a horse: it's not just about the right technique. You have to *feel* out how it works: how do you get the right balance? What is the rhythm of the horse? At what angle do you hold the bow and arrow? When do you aim? – It's about the process. And the horse tells you things you don't know." That's why he always wanted to go on a winter camp in the forest: "You learn how to get by when everything tumbles. And that's how you can teach people from the city that yearning for nature – and to experience that you don't have to live at a distance. And if you go to the forest, you can get by." His message to the children, who come to his camps, is therefore: "You too can become an Indian!" When he teaches them horseback riding, how to feed horses and communicate respectfully with them, how to live in the forest, it is about the process of attending and learning from others. And as an "Indian," he is something like an outsider, and the children usually listen to him more than to their teachers.

For Carsten, being "Indian", or the "Indian idea" as he calls it, is about thinking differently and about reminding yourself of the environment you live in, as well as the animals you live with. Being "Indian" is never a finished

19 See also Karl Marx, *Capital: Volume One* (London: Penguin Classics, 1999), 165.

process but rather a form of becoming that exceeds existing ways of acting and knowing.[20] But most of all it is about communication, about reconnecting to and re-enchanting the social landscape. Producing a "Wild West space," he has tried to repossess a sense of community in a world gone awry after the transformation to "capitalism," when everyone became stressed and hid in their homes, as he phrased it. In these face of the developments, what would be more appropriate than a community in wilderness? In a world shaped by the unruly social forces of a "wild capitalism,"[21] Carsten created what he saw as an oppositional space of unruliness in which he shifted between being computer expert and "Indian archer."

The Western Frontier in German Cultural Imaginations

The images of the "East" as wild space that Carsten's "powwows" as well as other enactments of Brandenburg's landscape as "wild," draw on and echo an old motive of the Eastern frontier: a spatial imaginary that engages longstanding German fascinations with the Wild West and that also played into Nazi ideas of the "wild, backward" East in need of colonization.[22] In Germany, these yearnings for "Indians," the American frontier and notions of "wilderness" have corresponded with various nationalist and imperial desires for expansion of the "homeland" throughout German history.[23] In addition, with the onset of urbanization and increasing industrialization,

20 For a discussion of "becoming" as analytical ethnographic lens that attends to emerging modes of existence and the ways in which their incompleteness exceeds prevalent modes of knowing or being in the world, see Joao Biehl and Peter Locke, *Unfinished. The Anthropology of Becoming* (Durham, NC: Duke University Press, 2017).

21 Krista Harper, *Wild Capitalism. Environmental Activism and Post-Socialist Ecology in Hungary* (Amherst: University of Massachusetts Press, 2006).

22 Larry Wolff. *Inventing Eastern Europe: The Map of Civilization on the Mind of the Enlightenment* (Stanford, CA: Stanford University Press, 1994).

23 Blackbourn, *Conquest of Nature*, 9.

many Germans began to identify with Native Americans as victims of environmental destruction, urbanization and the fragmentation of communities. In the twentieth century, Karl May stories and images of the American West inspired Nazi ideas of the "wild, backward" East in need of colonization. At the same time a "nostalgia for the wild" and the "noble savage" played a key role in Nazi racial discourse.[24] Thus, wilderness was both yearned for, and it had to be tamed and controlled.[25] Then, after the war, Karl May and the figure of the "Indian" continued to have a strong impact within German popular cultures, ranging from the so-called Winnetou movies in the mid-1960s in West Germany,[26] to the anti-capitalist Westerns in East Germany, where Karl May was frowned upon. At the same time, a large subculture of "Indian" and hobbyists emerged throughout both Germanies Eastern Europe.[27] Outdoor performances of Karl May stories, as well as Indian powwows became increasingly popular – not only among hobbyists, but also among thousands of Germans during the Cold War and they continue to capture large audiences today.[28]

Today, there are several Wild West towns and "Indian" villages in the region. In "Old Texas Town in Berlin," visitors can visit a gold mine, stop for a drink in the saloon, show their horse at the Blacksmith, exchange gold nuggets

24 Arnold Arluke, Boria Sax, "The Nazi Treatment of Animals and People," in Lynda Birke and Ruth Hubbard, eds., *Reinventing Biology. Respect for Life and the Creation of Knowledge* (Bloomington: Indiana University Press, 1995), 28–60.

25 For a discussion of American wilderness ethics and the ways in which German notions of *Landschaft* merged an idea of natural, built and cultivated environments at the basis of environmental care and a sense of home (*Heimat*) and national belonging, see Thomas Lekan, *Imagining the Nation in Nature: Landscape Preservation and German Identity 1885–1945* (Cambridge, MA: Harvard University Press, 2004), 15.

26 West German productions of the Karl May stories were immensely popular throughout Europe in the 1960s and 1970s.

27 Ruth Ellen Gruber, "Sauerkraut Cowboys and Klezmer Cafes," paper presentation, UC Santa Cruz, October 5, 2010. In addition, country music was imported to Europe and Germany in particular after World War II and thus incorporated even more into Wild West hobby cultures.

28 The so-called Karl May festival near Hamburg has been one of the most popular outdoor performances in Germany until today, attracting up to 300,000 visitors each year.

for money at the Bank of Texas, go square-dancing or stop by for a drink at the saloon.[29] In addition, idioms such as "Ein Indianer kennt kein Schmerz" [which translates into something like: "If you want to be an Indian, you have to be tough!"] have become an integral part of the vernacular throughout the twentieth century. In 2007, the newsmagazine *Der Spiegel* proclaimed Winnetou as the quintessential German national hero – including the longing for this figure, and the wilderness associated with it – as something as "German" as the myth of the forest, or German efficiency.[30]

In addition to this history of German and European fascinations with the "Wild West," cultural imaginaries of the March Oder region, of Brandenburg and "the East" as "untamed land" also have their own specific trajectory. David Blackbourn has traced the mystique of the frontier between wilderness and civilization expressed in narratives about the Wild East.[31] Frontier images emerged ever since Germans aspired to colonize eastern Prussia in the 1880s and imperial discourse, for example in the context of Frederick the Great, compared East European landscapes to North American ones.[32] Images of American westward expansion also played an important role in German discussions of colonial expansion and policymaking in South-West Africa.[33] Frontier imagery were also prevalent in nineteenth-century romanticist writing such as those of Theodor Fontane's *Rambles through the March Brandenburg* who compared Brandenburg's marshes with landscapes in the American West.[34] During World War II, the American frontier fed into fascist ideas of Germans as a people without land, in need of expansion.[35]

29 Old Texas Town was founded in Berlin in 1939 and, according to the town's website, had to soon "go underground" because of the political circumstances until 1950. See more on the history of the town at: <http://www.old-texas-town.de/>.
30 *Spiegel* magazine cited in Michael Kimmelman, "In Germany, Wild for Winnetou," *New York Times*. September 12, 2007.
31 See Blackbourn, *Conquest of Nature*.
32 Blackbourn, *Conquest of Nature*, 303.
33 Jens-Uwe Guettel. "From the Frontier to German South-West Africa: German Colonialism, Indians, and American Westward Expansion," *Modern Intellectual History* 7, no. 3 (2010), 523–552.
34 Blackbourn, *Conquest of Nature*, 293.
35 Blackbourn, *Conquest of Nature*, 293.

***

While Carsten's, and other recent enactments of Brandenburg as "wild" echo longer histories of German and European fascinations with the American West, they also articulate with interrupted temporalities and shifting cultural geographies after the Fall of the Wall. Producing a "wild West space" – the "Indian village," and the Western town attest to a change in affect to the postsocialist material environment,[36] while repossessing a sense of community in a world having gone astray after the transformation to capitalism. Although Carsten saw himself as something like an ambassador, he did not claim authenticity. Rather, as he told me in our conversations, he created a space of communication and connection to the environment, but also a space of unruliness in which you don't let yourself "be pinned down" by the social forces of the labor market – a space in which you shift between being computer expert and "Indian" archer. Yet ultimately, for Carsten "wilderness" is also a strategy in which stereotypes about the March Oder region as "backward" are transformed into an asset and establish a sense of regional identity at the same time.[37] Being "Indian," for Carsten, is about both reconnecting to and re-enchanting an imagined "natural" landscape and in this regard his performance of the local postsocialist landscape as "savage space" also caters to nostalgic (Western) conceptions of the socialist past.[38]

Yet most importantly, these postsocialist yearnings and affective landscapes mingle with an ambivalent and deeply fraught desire for racialized

36 For an analysis of post/socialist affect and the urban material environment in Vietnam, see Christina Schwenkel, "Post/Socialist Affect: Ruination and Reconstruction of the Nation in Urban Vietnam," *Cultural Anthropology* 28, no. 2 (2013), 252–277, as well as Schwenkel's contribution to this volume.

37 Heatherington develops a similar argument in *Wild Sardinia*, 303: while rural Sardinians authenticate and ironize the notion of the backwardness of the central highlands, they seize nationalist tropes of time. Embracing "barbarian" ecological ideas and invoking the figure of the primitive, they also mock global perspectives of modern environmentalism and conservation. While thus entering the political field, they nevertheless reinforce stereotypes of the Sardinian wilderness.

38 On "Ostalgie," see also Dominic Boyer, "Ostalgie and the Politics of the Future in Eastern Germany," *Public Culture* 18, no. 2 (2006), 361–381.

otherness.[39] While Carsten's practice thus integrated contradictory agendas to articulate a critique of capitalism, his critique was routed through a discourse of race and an enactment of difference. Mimicking the colonized other has been a longstanding tradition that served white European self-articulations throughout the history of colonial encounter, and is deeply connected to histories of material exploitation.[40] Today, such mimicry easily enacts a form of whiteness that does not have to position itself vis a vis engrained colonial legacies[41] and cultural appropriations of indigeneity – and that, as indigenous scholars and activists have pointed out for a long time, ignores past harm and evades the complexity of ongoing indigenous struggles for sovereignty and integrity.[42]

Yet if these postsocialist renderings of Brandenburg's wilderness offered a deeply fraught space for working through the economic, social and ecological ruptures of Germany's transition, contemporary European

39 Kathrin Sieg's work has been central in showing that "Indian" re-enactments are heterogeneous cultural practices that not simply echo older racial discourses but integrate antiracist, anti-materialist, environmentalist cultural agendas with racist ones; see Kathrin Sieg, *Ethnic Drag. Performing Race, Nation, Sexuality in West Germany* (Ann Arbor: University of Michigan Press, 2004). See also essays in Gerd Gemunden, Colin G. Calloway, Susanne Zantop, *Germans and Indians: Fantasies, Encounters, Projections*, University of Nebraska Press, 2002); and for a discussion of white American Indian reenactments and their significance in formulating American national identity, see Philipp Deloria, *Playing Indian* (London: Yale University Press, 1999). Most importantly, for critiques by indigenous scholars, see for example Audra Simpson, *Mohawk Interruptus. Political Life Across the Border of Settler States* (Durham, NC: Duke University Press, 2015) and more recent indigenous efforts to create accountability against cultural appropriation of indigenous cultures, <https://www.huffingtonpost.com/entry/un-cultural-appropriation-illegal_us_594073d3e4b0d3185485ba9a>.

40 bell hooks, *black looks: Race and Representation* (Boston, MA: South End Press, 1992), 182.

41 For a discussion on critical perspectives of whiteness in the German and European context, see Susan Arndt, Maureen Maisha Eggers, Grada Kilomba, Peggy Piesche, eds., *Mythen, Masken und Subjekte: Kritische Weißseinsforschung in Deutschland* (Münster: Unrast Verlag, 2017).

42 Simpson, *Mohawk Interruptus*. See also Eve Tuck and Wayne Yan,"Decolonization is not a Metaphor," *Decolonization: Indigineity, Education and Society* 1, no. 1 (2012), 1–40.

racial formations and the legacies of colonialism also found their way into local forests via a different route. Recent postcolonial displacements and migratory movements have restructured the European political and physical landscape. And in the process, the March Oder region has also been transformed as refugees from Africa and the Middle East have sought out Europe in search for a life away from war, political conflict and economic precarity. Let me therefore now turn to a different set of encounters with "wilderness" at the edges of Berlin – this time seen from the perspective of East African asylum seekers living in Brandenburg's forests.

The "Wilderness" of Contemporary Asylum Worlds

In the past two decades, the March Oder region has also become a site of debate over migration policies and racial exclusions. Long before the most recent debates about the 2015 "refugee crisis"[43] many asylum seekers – particularly from East Africa, but also Asia and the Middle East – have lived in Brandenburg's forests for years. This has been the consequence of post-unification asylum policies that created new urban shelters in East German cities in order to equally "re-distribute" asylum seekers in East and West. The 1990s were marked by heated nation-wide asylum debates in the aftermath of xenophobic attacks against urban refugee shelters after the war in former Yugoslavia. And it was in this context that many counties relocated refugees to new "homes" (*Heime*) in rural East Germany – often on the ruins of former military barracks in the forest. These "homes" have mostly been run by private contractors that signed long-term contracts with municipalities to accommodate asylum seekers. In the past decade, official legislation has for the most part not allowed refugees to work while asylum cases are pending; and the so-called residential law restricted travel outside

43 For a critical discussion of the crisis rhetoric around migration to Europe and Germany, see Seth Holmes, Heide Castaneda, "Representing the European Refugee Crisis in Germany and Beyond," *American Ethnologist* 43, no. 1 (2016): 12–24.

of the limits of the county in which asylum seekers are registered – which has made it illegal for many to take a train to Berlin, just a 35-kilometer ride away.[44] In addition, with asylum infrastructures scaled down significantly until the most recent reforms, many refugees have remained in legal limbo for sometimes up to eight or more years.

In the March Oder Park, a series of scandals has surrounded the location and relocation of several refugee camps on former military sites. One scandal involved a private subcontractor who ran a refugee accommodation near the town of Ronersdorf and illegally hired refugees living in the "home"/Heim for 1 Euro per hour to build an adjacent ostrich farm. Today, the ostrich farm is a popular tourist stop, advertised as the perfect destination for a weekend outing with "exotic" flair. When protests uncovered these exploitative practices, this and several other refugee camps were moved to other, often even more remote locations on former military grounds in the forest. Although some of these "homes" were announced to be only interim solutions, people often ended up living in them for many years.

Refugee activists and advocates have protested these out-of-the-way camps for years: criticizing their spatial isolation in addition to restriction of free movement by the residency law. Some activists have also argued that Germany's policies (and especially the residency law, and the distribution of people to Germany's peripheries) resemble strategies in Germany's former colonies that segregated native from European populations and surveilled their residency.[45] Indeed, such colonial forms of surveillance haunted many of the refugees living in the forest in the March Oder region. Many of them from West and East Africa, they had experienced the legacies of colonial rule – and these legacies now shaped their perceptions of the local environment in turn.

44 Due to protests by human rights organizations and recent asylum policy reform, some German states loosened and partially abolished the residency law in the last few years, <http://www.buzer.de/gesetz/6406/al63502-0.htm>. Yet in the aftermath of the debates about the "refugee crisis" in 2015, new voices have emerged to more strictly enforce regulations again in different states and on a federal level.

45 See: <http://thevoiceforum.org/node/308> and Njoh 1997 on Germany's colonial spatial policies.

Stories of unruly nature and the wild punctuated refugees' accounts of their surroundings and their everyday life in the region. Living on a converted military base in the forest near the small town of Zarin, Nangila had been in legal limbo and had not been allowed to work for many years.[46] He and other asylum seekers narrated their time in Germany as a time of being stuck in the "German bush." With these references to the bush, they described their lives as a space of wilderness, both in physical, symbolical and affective terms. For them, this was not the Wild East or a romantic landscape of exploration but a land gone awry. Whereas Europe stood for progress and was imagined as urban, the Brandenburg landscape stood in stark contrast to that image. Indeed, as their lives did not easily find traction, Brandenburg seemed less like Germany, or Europe to them, but more like another Kenya, or Cameroon, as one of my informants pointed out.

Many refugees thus recounted stories of living unruly lives in the "German bush." Brandenburg was a challenging place for them because, here, they sat around all day, unable to work, waiting for their asylum process to go through. Struggling to make a life in the "bush" meant having to come up with creative ideas to spend your time and overcome boredom. Often it simply meant sleeping a lot, going for a walk, or not being able to interact with anyone except with wild animals in the forest. One resident described encounters with strange, unknown beasts in the forest at night (that turned out to be wild boar). Another resident from Kenya, Maria, walked an hour through the forest each morning and night, often with a flashlight. In addition, the specter of violence was very prevalent in these stories: many refugees worried about or actually had been harassed by hostile local youth. This German "bush" was thus a forest quite different from the "untouched nature" praised in the March Oder Park. It was a "bush" in the Brandenburg "wilderness" in which one was vulnerable to isolation and unpredictable forces of potential violence. This violence was both physical and psychological. Many men had had encounters with Neo-Nazis in the region. Moreover, with local police regularly conducting "passport controls," being checked by the police was a constant risk. These stories about

46 To protect the anonymity of my informants, I use pseudonyms for both people and specific places here.

the German forest thus invoked affects and environments that captured the visceral dimensions of racialized exclusions emerging in contemporary asylum worlds in Germany – as well as their material effects.[47]

Conclusion

In this essay, I have traced different practices of narrating and performing the Brandenburg landscape as a space of wilderness. As local economies collapsed in the aftermath of unification and the region's environment has been undergoing profound change, the very category of unruly nature gains traction across different realms: military zones transform into nature spaces, formerly extinct "wild animals" re-enter the region, local tourist agencies advertise the region as an undiscovered land of ecological diversity, ready to be explored, and computer specialists pose as "Indians." Meanwhile, people fleeing war and political conflict across Africa and the Middle East move into former military spaces and, surrounded by unknown ecologies of the forest, encounter everyday isolation, uncertainty and the specter of racial violence.

Whereas people like Carsten formulated a critique of a capitalism gone awry via their enactment of wild Brandenburg, the refugee stories about the Brandenburg "bush" offer a different and in fact central angle onto the postsocialist landscape in Brandenburg. In their references to the German "bush," Nangila and Maria formulated critical commentaries on the violence and "wildness" of contemporary European asylum worlds and the exclusions that emerge in their midst. From the perspective of these

47 For a more detailed discussion of an ethnographic focus on material environments and the embodied dimensions of racialized exclusions – as well as its possibilities for departing from dominant conventions of narrating the "true lives" of immigrants in contemporary asylum worlds, see Bettina Stoetzer, "A Path Through the Woods: Remediating Affective Landscapes in Contemporary Asylum Worlds," *Transit* 9, no. 2 (2014).

refugees, the "wild" constitutes a space in which one's psychological and physical integrity is at risk. The wild landscape many asylum seekers referred to was the effect of a process with a long colonial history of making "wild country"[48] – a racialized process of being placed within "nature," a space of degradation and unruliness over which Europeans have the power of administration. This space was marked by a constant sense of precarity in the face of illogical bureaucratic regulations, seemingly arbitrary laws, the risk of racial violence and a daily struggle of fighting boredom or despair – as well as long walks through the forest. With their stories, these refugees thus situated East German narratives and performances of wilderness, such as Carsten's, in a decidedly postcolonial context. Doing so, they shed light on the urgency of attending to the ways in which shifting racial formations materialize in contemporary Europe's postsocialist environments.

48 Deborah Bird Rose, *Reports from a Wild Country. Ethics for Decolonization* (Sydney: University of New South Wales Press, 2004).

Notes on Contributors

JULIA E. AULT is Assistant Professor of History at the University of Utah. She completed her PhD at the University of North Carolina at Chapel Hill in 2015. Her first book project explores East German environmental policy and protest in a transnational, Central European context, with a focus on connections to West Germany and Poland. She has received numerous fellowships, including from the Central European History Society, the Free University's Berlin Program, and Columbia University's Council for European Studies. She recently published "Defending God's Creation: The Environment in State, Church and Society in the GDR, 1975–89" in *German History* and has a chapter in John R. McNeill and Astrid Mignon Kirchhof's *Nature and the Iron Curtain: Environmental Policy and Social Movements in Communist and Capitalist Countries, 1945–1990* (University of Pittsburgh Press). Her second book-length project explores the importance of water and water management to environmental policy in the "second world."

MICHEL DUPUY is an associate researcher at the Institute of Modern and Contemporary History in Paris. After his PhD on the history of forest ecology in France and in Germany (1880–1980), he worked at the Centre Marc Bloch in Berlin, where he decided to investigate the environmental history of the GDR through different contexts, including scientific discourse on air pollution, especially in the public arena; hunting; forest recreation; and global warming. For 15 years, he has studied the media treatment of environmental issues on French television, with a focus on scientific discourse.

THOMAS FLEISCHMAN is Assistant Professor of German History at the University of Rochester. He is the author of *Three Little Pigs: An Animal History of East Germany, 1945–2014*, which is forthcoming with Weyerha euser Environmental Books at the University of Washington Press. *Three Little Pigs* re-examines the agricultural pollution epidemic that beset the

"Workers' and Peasants' State" in the 1980s, locating its origins in the emergent nexus between communist agriculture and capitalist trade during the Cold War. Beyond agriculture, he is interested in the variety of roles animals play in human history, and how "thinking" with living creatures reshapes our understanding of the past. To that end, he has begun a new research project about whales, whaling, and German history from approximately 1885 to 1960.

TOBIAS HUFF (Dr. phil.) is Director of the Faculty of History and Cultural Studies at the University of Mainz. He studied history, economics and political science in Mainz, Zurich and Gävle, Sweden. His MA thesis on the social impact of the Tambora eruption in 1815 has won several awards. Afterwards he completed his doctorate in Freiburg in economic and social history on the environmental policy of the GDR. He then worked as a PR editor in the field of renewable energy before returning to the university. He researches and publishes on issues of early modern economic history, risk history and environmental history.

ASTRID MIGNON KIRCHHOF is a senior researcher and lecturer at Humboldt University, Berlin. In 2016–19 she was research associate and scholar-in-residence at the Deutsches Museum, Munich in the EU-funded project History of Nuclear Energy and Society. Previously she was a Volkswagen Foundation Postdoctoral Fellow in the Humanities at Georgetown University and the German Historical Institute, Washington, DC. From 2010 to 2014 she was principal investigator of a research project about the nature conservation movement in East and West Berlin between 1945 and 1990, funded by the German Research Foundation at the Humboldt University. Her co-edited book (with John R. McNeill), *Nature and the Iron Curtain. Environmental Policy and Social Movements in Communist and Capitalist Countries, 1945–1990*, was published by University of Pittsburgh Press in 2019.

SABINE MÖDERSHEIM is Professor of German Literature and Culture and Director of the Center for Early Modern Studies at the University of Wisconsin-Madison. Her research interests are in the areas of literature and culture from the Renaissance to the present, in particular material and visual culture, the use of images in architectural decorations, popular culture and propaganda, as well as German environmental history and its media. Current research projects in ecocriticism and environmental study focus on the history of environmental grass-roots movements in Germany, as well as environmental policies and best practices in urban planning. She serves on the editorial board of Monatshefte für deutschsprachige Literatur und Kultur. Recent publications include *Deutsche Geheimgesellschaften von der Frühen Neuzeit bis zur Gegenwart*, co-edited with Jost Hermand, and *The Art of Persuasion: Emblems and Propaganda*.

SCOTT MORANDA is Professor of History at SUNY Cortland, where he teaches environmental, German, and central European history. He published *The People's Own Landscape: Nature, Tourism, and Dictatorship in East Germany* in 2014 with the University of Michigan Press. He is currently working on an environmental history of the movement of German-speakers back and forth across the Atlantic between 1860 and 1960. This project will trace the transatlantic movement of ideas about soil and forest conservation, but also the migration of biota enabled by these migrations. His work, more generally, explores the significance of tourism, travel, and migration in environmental history.

KATRINA NOUSEK's research currently focuses on narrative structures of temporality in contemporary literature about globalization. Her wider teaching and research interests include visual culture, environmental humanities, and narrative studies. She is currently working on *The Vanishing Point of the Future*, a book that analyzes futurity in literary migration narratives addressing postsocialism in Europe. She holds a PhD in German Studies from Cornell University and a BA in History and Literature from Harvard University. She is Visiting Assistant Professor of German Studies at the University Richmond.

ELI RUBIN is the author of two monographs: *Synthetic Socialism: Plastics and Dictatorship in the German Democratic Republic* (University of North Carolina Press, 2008) and *Amnesiopolis: Modernity, Space, and*

Memory in East Germany (Oxford University Press, 2016). He was a Humboldt Postdoc from 2007–9 and a recipient of grants from the German Historical Institute and the DAAD. He is currently working on a third monograph about destruction and urban space in Germany during World War II. He is Professor of History at Western Michigan University, and received his PhD from the University of Wisconsin-Madison in 2004.

CHRISTINA SCHWENKEL is Associate Professor of Anthropology at the University of California, Riverside and Director of the Program in Southeast Asian Studies (SEATRiP). She is co-editor-in-chief of the *Journal of Vietnamese Studies*. Her book, *The American War in Contemporary Vietnam: Transnational Remembrance and Representation* (Indiana University Press, 2009), examines historical knowledge production and the geopolitics of postwar commemoration. She has conducted extensive ethnographic research in Vietnam and eastern Germany. Her publications have focused on Cold War visuality, socialist mobilities and the making of expertise, and reconstruction of urban infrastructure following the end of US air strikes in Vietnam. Her book on East German architectural transfers to the global South and the ruins of utopias past in urban Vietnam is forthcoming.

BETTINA STOETZER is a cultural anthropologist and associate professor at MIT. Her research focuses on the intersections of ecology, migration, and urban social justice. Her forthcoming book *Ruderal City: Ecologies of Migration and Urban Life in Berlin* (Duke University Press) examines how human-environment relations become a key register through which urban citizenship is articulated in contemporary Europe. Bettina Stoetzer is also the author of *InDifferenzen: Feministische Theorie in der Antirassistischen Kritik* (argument, 2004), *Shock and Awe. War on Words*, together with Gonzalez, Eekelen and Tsing (New Pacific Press, 2004), along with articles on urban ecology, race, affect, and migration. She is currently working on a new project on wildlife mobility, climate change, and nationalism in the US and Germany.

GERNOT WALDNER studied Literature, Philosophy, and the History of Science in Vienna and Berlin. He received his PhD in German Studies from Harvard University in 2018. His dissertation analyzes the usage of statistical models in early twentieth-century prose (Broch, Musil, Neurath, Frank). He has published articles on the impact of Cybernetics on Stanley Kubrick's *2001: A Space Odyssey*, the connection between grotesque humor and racial theories in the writings of Fritz von Herzmanofsky Orlando and several topics related to Red Vienna. He has translated books and articles from English to German, from Italian to German and from German to English. He has curated exhibitions on Red Vienna and on Otto Neurath's statistical picture language ISOTYPE. His current research is concerned with the transnational history of Alfred Adler's psychology during the interwar period, how a psychological school that shaped the policies of socialist-governed Vienna become a popular school of psychology in the United States.

Index

German Life and Civilization

German Life and Civilization provides contributions to a critical understanding of Central European cultural history from medieval times to the present. Culture is here defined in the broadest sense, comprising expressions of high culture in such areas as literature, music, pictorial arts, and intellectual trends as well as political and sociohistorical developments and the texture of everyday life. Both the cultural mainstream and oppositional or minority viewpoints lie within the purview of the series. While it is based on specialized investigations of particular topics, the series aims to foster progressive scholarship that aspires to a synthetic view of culture by crossing traditional disciplinary boundaries.

1 Charlotte L. Brancaforte: The German Forty-Eighters in the United States. 1989. US-ISBN 0-8204-1010-1.

2 Heinz D. Osterle (ed.): Amerika: New Images in German Literature. 1989. US-ISBN 0-8204-1058-6.

3 Luke Springman: Comrades, Friends and Companions. Utopian Projections and Social Action in German Literature for Young People 1926–1934. 1989. US-ISBN 0-8204-0963-4.

4 Peter Morris-Keitel: Die Verbrechensthematik im modernen Roman. 1989. US-ISBN 0-8204-1106-X.

5 Sussan Milantchi Ameri: Die deutschnationale Sprachbewegung im Wilhelminischen Reich. 1991. US-ISBN 0-8204-1119-1.

6 Michael Myers: Für den Bürger. The Role of Christian Schubart's "Deutsche Chronik" in the Development of a Political Public Sphere. 1990. US-ISBN 0-8204-1168-X.

7 James W. Jones: "We of the Third Sex". Literary Representations of Homosexuality in Wilhelmine Germany. 1990. US-ISBN 0-8204-1209-0.

8 Rachel J. Halverson: Historiography and Fiction. Siegfried Lenz and the "Historikerstreit". 1991. US-ISBN 0-8204-1288-0.

9 Kirsten Belgum: Interior Meaning. Design of the Bourgeois Home in the Realist Novel. 1991. US-ISBN 0-8204-1627-4.

10 Barbara Völkel: Karl Philipp Moritz und Jean-Jacques Rousseau. Außenseiter der Aufklärung. 1991. US-ISBN 0-8204-1689-4.

11 Dagmar C. G. Lorenz: Verfolgung bis zum Massenmord. Holocaust-Diskurse in deutscher Sprache aus der Sicht der Verfolgten. 1992. US-ISBN 0-8204-1751-3.

12 Alexander Mathaes: Der Kalte Krieg in der deutschen Literaturkritik. Der Fall Martin Walser. 1992. US-ISBN 0-8204-1824-2.

13 Amy Stapleton: Utopias for a Dying World. Contemporary German Science Fiction's Plea for a New Ecological Awareness. 1993. US-ISBN 0-8204-1922-2.

14 K. Schultz, K. Calhoon (eds.): The Idea of the Forest. German and American Perspectives on the Culture and Politics of Trees. 1996. US-ISBN 0-8204-2384-X.

16 Jost Hermand (ed.): Postmodern Pluralism and Concepts of Totality. The Twenty-fourth Wisconsin Workshop. 1995. US-ISBN 0-8204-2658-X.

17 Myra J. Heerspink Scholz: A Merchant's Wife on Knight's Adventure. Permutations of a Medieval Tale in German, Dutch, and English Chap-books around 1500. 1999. US-ISBN 0-8204-2573-7.

18 H. Adler, J. Hermand (eds.): Günter Grass. Ästhetik des Engagements. 1996. US-ISBN 0-8204-2719-5.

19 Christian Essellen: Babylon. Edited and with an Introduction by Cora Lee Nollendorfs. 1996. US-ISBN 0-8204-3045-5.

20 Klaus L. Berghahn (ed.): The German-Jewish Dialogue Reconsidered. A Symposium in Honor of George L. Mosse. 1996. US-ISBN 0-8204-3107-9.

21 Andréa Staskowski: Conversations with Experience. Feminist Hermeneutics and the Autobiographical Films of German Women. 2004. US-ISBN 0-8204-3128-1.

22 J. Hermand, J. Steakley (eds.): "Heimat", Nation, Fatherland. The Ger-man Sense of Belonging. 1996. US-ISBN 0-8204-3373-X.

23 Jörg Bernig: Eingekesselt. Die Schlacht um Stalingrad im deutschsprachigen Roman nach 1945. 1997. US-ISBN 0-8204-3667-4.

24 J. Hermand, M. Silberman (eds.): Contentious Memories. Looking Back at the GDR. 251 pp. 1998, 2000. US-ISBN 0-8204-5254-8.

25 Franziska Meyer: Avantgarde im Hinterland. Caroline Schlegel-Schelling in der DDR-Literatur. 1999. US-ISBN 0-8204-3924-X.

26 J. Hermand, R. Holub (eds.): Heinrich Heine's Contested Identities. Politics, Religion, and Nationalism in Nineteenth-Century Germany. 1999. US-ISBN 0-8204-4105-8.

27 Linda F. McGreevy: Bitter Witness. Otto Dix and the Great War. 467 pp. 2001. US-ISBN 0-8204-4106-6.

28 H. Adler, J. Hermand (eds.): Concepts of Culture. 1999. US-ISBN 0-8204-4141-4.

29 Hannu Salmi: Imagined Germany. Richard Wagner's National Utopia. 1999. US-ISBN 0-8204-4416-2.

30 Dennis Brain: Johann Karl Wezel. From Religious Pessimism to Anthro-pological Skepticism. 1999. US-ISBN 0-8204-4464-2.

31 Karoline von Oppen: The Role of the Writer and the Press in the Unification of Germany, 1989–1990. 277 pp. 2000. US-ISBN 0-8204-4488-X.

32 J. Hermand, M. Silberman (eds.): Rethinking Peter Weiss. 199 pp. 2000, 2001. US-ISBN 0-8204-5819-8.

33 Peter Morris-Keitel, Michael Niedermeier (eds.): Ökologie und Literatur. 224 pp. 2000. US-ISBN 0-8204-4872-9.

34 Helmut Peitsch: Georg Forster. A History of His Critical Reception. 333 pp. 2001. US-ISBN 0-8204-4925-3.

35 Martin Travers: Critics of Modernity. The Literature of the Conservative Revolution in Germany, 1890–1933. 256 pp. 2001. US-ISBN 0-8204-4927-X.

36 Susan C. Anderson: Water, Culture and Politics in Germany and the American West. 310 pp. 2001. US-ISBN 0-8204-4946-6.

37 Katherine Arens: Empire in Decline. Fritz Mauthner's Critique of Wilhelminian Germany. 222 pp. 2001. US-ISBN 0-8204-5038-3.

38 K. Berghahn, J. Fohrmann, H. Schneider (eds.): Kulturelle Repräsentationen des Holocaust in Deutschland und den Vereinigten Staaten. 253 pp. 2002. US-ISBN 0-8204-5208-4.

39 Hans Adler (ed.): Aesthetics and Aisthesis. New Perspectives and (Re)Discoveries. 170 pp. 2002. ISBN 3-906768-40-6, US-ISBN 0-8204-5852-X.

40 Silke von der Emde: Entering History. Feminist Dialogues in Irmtraud Morgner's Prose. 260 pp. 2004. ISBN 3-03910-158-7, US-ISBN 0-8204-6968-8.

41 Jost Hermand: Pro und Contra Goethe. Dichterische und germanistische Stellungnahmen zu seinen Werken. 214 pp. 2005.
ISBN 3-03910-175-7, US-ISBN 0-8204-6985-8.

42 Bill Maltarich: Samurai and Supermen. National Socialist Views of Japan. 406 pp. 2005.
ISBN 3-03910-303-2, US-ISBN 0-8204-7228-X.

43 Eric Paul Jacobsen: From Cosmology to Ecology. The Monist World-View in Germany from 1770 to 1930. 401 pp. 2005.
ISBN 3-03910-306-7, US-ISBN 0-8204-7231-X.

44 Klaus L. Berghahn, Jost Hermand (eds.): Unmasking Hitler. Cultural Representations of Hitler from the Weimar Republic to the Present. 300 pp. 2005. ISBN 3-03910-553-1, US-ISBN 0-8204-7581-5.

45 Jost Hermand (ed.), Positive Dialektik. Hoffnungsvolle Momente in der deutschen Kultur. Festschrift für Klaus L. Berghahn zum 70. Geburtstag. 304 pp. 2007. ISBN 978-3-03-910946-3.

46 Matthew Lange: Antisemitic Elements in the Critique of Capitalism in German Culture, 1850–1933.
348 pp. 2007. ISBN 978-3-03911-040-7.

47 Helen Fehervary & Bernd Fischer (Hrsg./eds): Kulturpolitik und Politik der Kultur/Cultural Politics and the Politics of Culture. Festschrift für Alexander Stephan/Essays to Honor Alexander Stephan.
489 pp. 2007. ISBN 978-3-03911-076-6.

48 Bryan Ganaway: Toys, Consumption, and Middle-class Childhood in Imperial Germany, 1871–1918.
299 pp. 2009. ISBN 978-3-03911-548-8.

49 Corina L. Petrescu: Against All Odds. Models of Subversive Spaces in National Socialist Germany. 284 pp. 2010. ISBN 978-3-03911-845-8.

50 Jerry Schuchalter: Poetry and Truth. Variations on Holocaust Testimony. 302 pp. 2009. ISBN 978-3-03911-857-1.

51 Jost Hermand: Fünfzig Jahre Germanistik. Aufsätze, Statements, Polemiken 1959–2009. 402 pp. 2009. ISBN 978-3-03911-877-9.

52 Babette Babich: Nietzsches Wissenschaftsphilosophie. «Die Wissenschaft unter der Optik des Künstlers zu sehn, die Kunst aber unter der des Lebens». 405 pp. 2011. ISBN 978-3-03911-945-5.

53 Weijia Li: China und China-Erfahrung in Leben und Werk von Anna Seghers. 269 pp. 2010. ISBN 978-3-0343-0157-2.

54 Cora Lee Kluge (ed.): Paths Crossing. Essays in German-American Studies. 191 pages. 2011. ISBN 978-3-0343-0221-0.

55 Russell T. Harrison: Thomas Bernhard's Comic Materialism. Class, Art, and "Socialism" in Post-War Austria. 206 pp. 2012. ISBN 978-3-0343-0286-9.

56 Jost Hermand: Unerfüllte Hoffnungen. Rückblicke auf die Literatur der DDR. 304 pages. 2012. ISBN 978-3-0343-0886-1.

57 Alan Corkhill: Spaces for Happiness in the Twentieth-Century German Novel. Mann, Kafka, Hesse, Jünger. 211 pages. 2012. ISBN 978-3-0343-0797-0.

58 Shafiq Shamel: Goethe and Hafiz. Poetry and History in the *West-östlicher Divan*. 273 pages. 2013. ISBN 978-3-0343-0881-6.

59 Jochen Vogt: Erinnerung, Schuld und Neubeginn. Deutsche Literatur im Schatten von Weltkrieg und Holocaust. 451 pages. 2014. ISBN 978-3-0343-0855-7.

60 Sylvia Fischer: *Dass Hämmer und Herzen synchron erschallen*. Erkundungen zu Heimat in Literatur und Film der DDR der 50er und 60er Jahre. 287 pages. 2015. ISBN 978-3-0343-1877-8.

61 Lesley-Ann Brown: The German *Lied* after Hugo Wolf. From Hans Pfitzner to Anton Webern. 298 pages. 2015. ISBN 978-3-0343-1888-4.

62 Andrea Fieler: W.G. Sebalds *Nach der Natur*. Eine Analyse. 188 pages. 2015. ISBN 978-3-0343-0292-0.

63 Marc J. Schweissinger: Vom bürgerlichen zum sozialen Trauerspiel Gerhart Hauptmanns. 240 pages. 2016. ISBN 978-1-906165-76-5.

64 Hilary Potter: Remembering Rosenstrasse. History, Memory and Identity in Contemporary Germany. Forthcoming. ISBN 978-3-0343-1917-1.

65 Jessica Macauley: Forces of Ambiguity. Life, Death, Disease and Eros in Thomas Mann's *Der Zauberberg*. 330 pages. 2017. ISBN 978-1-78707-237-4.

66 Edward T. Larkin: Carl Wilhelm Frölich's *On Man and his Circumstances*. A Translation of *Über den Menschen und seine Verhältnisse*. 272 pages. 2017. ISBN 978-1-78707-311-1.

67 Francesca Goll: Mapping Spaces. Reimagining East German Society in 1960s Fiction. 262 pages. 2019. ISBN 978-1-78707-917-5.

68 Marc Silberman (ed.): Back to the Future. Tradition and Innovation in German Studies. 278 pages. 2018. ISBN 978-1-78874-305-1.

69 William J. Waltz: Of Writers and Workers. The Movement of Writing Workers in East Germany. 264 pages. 2018. ISBN 978-1-78874-499-7.

70 Sabine Mödersheim, Scott Moranda and Eli Rubin (eds): Ecologies of Socialisms. Germany, Nature, and the Left in History, Politics, and Culture. 346 pages. 2019. ISBN 978-1-78707-577-1.

www.ingramcontent.com/pod-product-compliance
Ingram Content Group UK Ltd.
Pitfield, Milton Keynes, MK11 3LW, UK
UKHW021853100926
8556IPUK00020B/1197